SCIENCE

A CLOSER LOOK

Macmillan McGraw-Hill

Program Authors

Dr. Jay K. Hackett
Professor Emeritus of Earth Sciences
University of Northern Colorado
Greeley, CO

Dr. Richard H. Moyer
Professor of Science Education and Natural Sciences
University of Michigan-Dearborn
Dearborn, MI

Dr. JoAnne Vasquez
Elementary Science Education Consultant
NSTA Past President
Member, National Science Board and NASA Education Board

Mulugheta Teferi, M.A.
Principal, Gateway Middle School
Center of Math, Science, and Technology
St. Louis Public Schools
St. Louis, MO

Dinah Zike, M.Ed.
Dinah Might Adventures LP
San Antonio, TX

Kathryn LeRoy, M.S.
Executive Director
Division of Mathematics and Science Education
Miami-Dade County Public Schools, FL
Miami, FL

Dr. Dorothy J. T. Terman
Science Curriculum Development Consultant
Former K-12 Science and Mathematics Coordinator
Irvine Unified School District, CA
Irvine, CA

Dr. Gerald F. Wheeler
Executive Director
National Science Teachers Association

Bank Street College of Education
New York, NY

Contributing Authors

Dr. Sally Ride
Sally Ride Science
San Diego, CA

Lucille Villegas Barrera, M.Ed.
Elementary Science Supervisor
Houston Independent School District
Houston, TX

American Museum of Natural History
New York, NY

Contributing Writer

Ellen C. Grace, M.S.
Consultant
Albuquerque, NM

RFB&D learning through listening Students with print disabilities may be eligible to obtain an accessible, audio version of the pupil edition of this textbook. Please call Recording for the Blind & Dyslexic at 1-800-221-4792 for complete information.

C

The McGraw-Hill Companies

Macmillan/McGraw-Hill

Send all inquiries to:
Macmillan/McGraw-Hill
8787 Orion Place
Columbus, OH 43240-4027

FOLDABLES™ is a Trademark of The McGraw-Hill Companies, Inc.

ISBN: 978-0-02-284134-8
ISBN: 0-02-284134-2

Printed in the United States of America.

14 15 16 17 LWI 23 22 21 20

The American Museum of Natural History in New York City is one of the world's preeminent scientific, educational, and cultural institutions, with a global mission to explore and interpret human cultures and the natural world through scientific research, education, and exhibitions. Each year the Museum welcomes around four million visitors, including 500,000 schoolchildren in organized field trips. It provides professional development activities for thousands of teachers; hundreds of public programs that serve audiences ranging from preschoolers to seniors; and an array of learning and teaching resources for use in homes, schools, and community-based settings. Visit www.amnh.org for online resources.

Be a Scientist

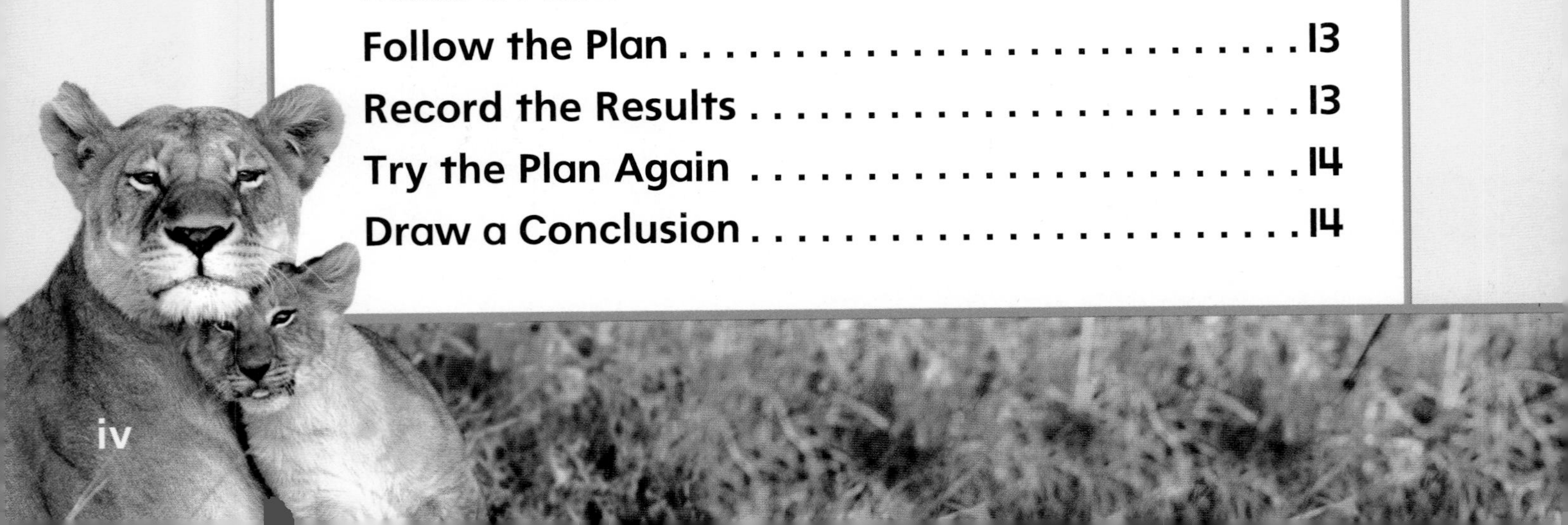

Scientific Method

1. Observe
2. Ask a Question
3. Make a Prediction
4. Make a Plan
5. Follow the Plan
6. Record the Results
7. Try the Plan Again
8. Draw a Conclusion

Life Science

UNIT A Plants

UNIT B Animals and Their Homes

Earth Science

UNIT C Our Earth

UNIT D Weather and Sky

Physical Science

UNIT E Matter

UNIT F Motion and Energy

Activities and Investigations

Life Science

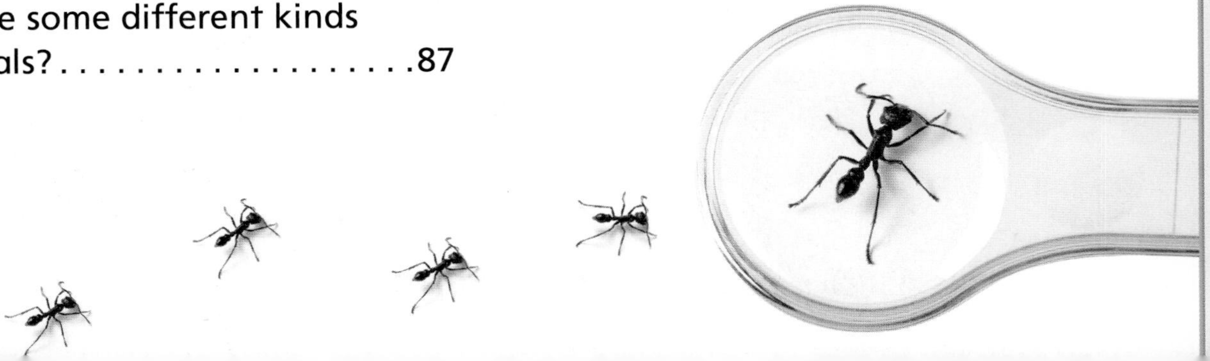

Be a Scientist investigations and Focus on Skills skill builders are found in the Activities Flipchart.

Earth Science

Activities and Investigations

Physical Science

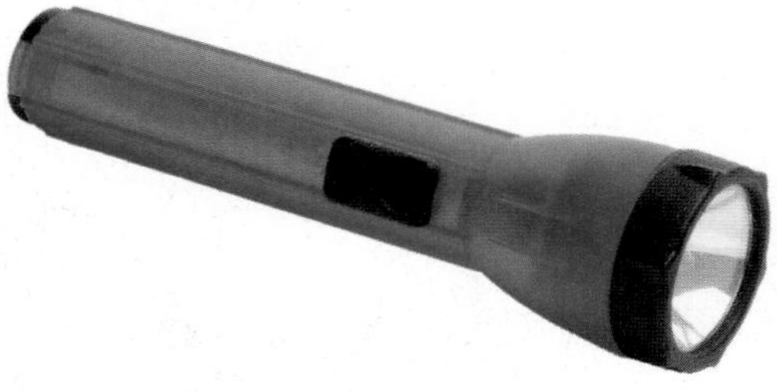

Be a Scientist

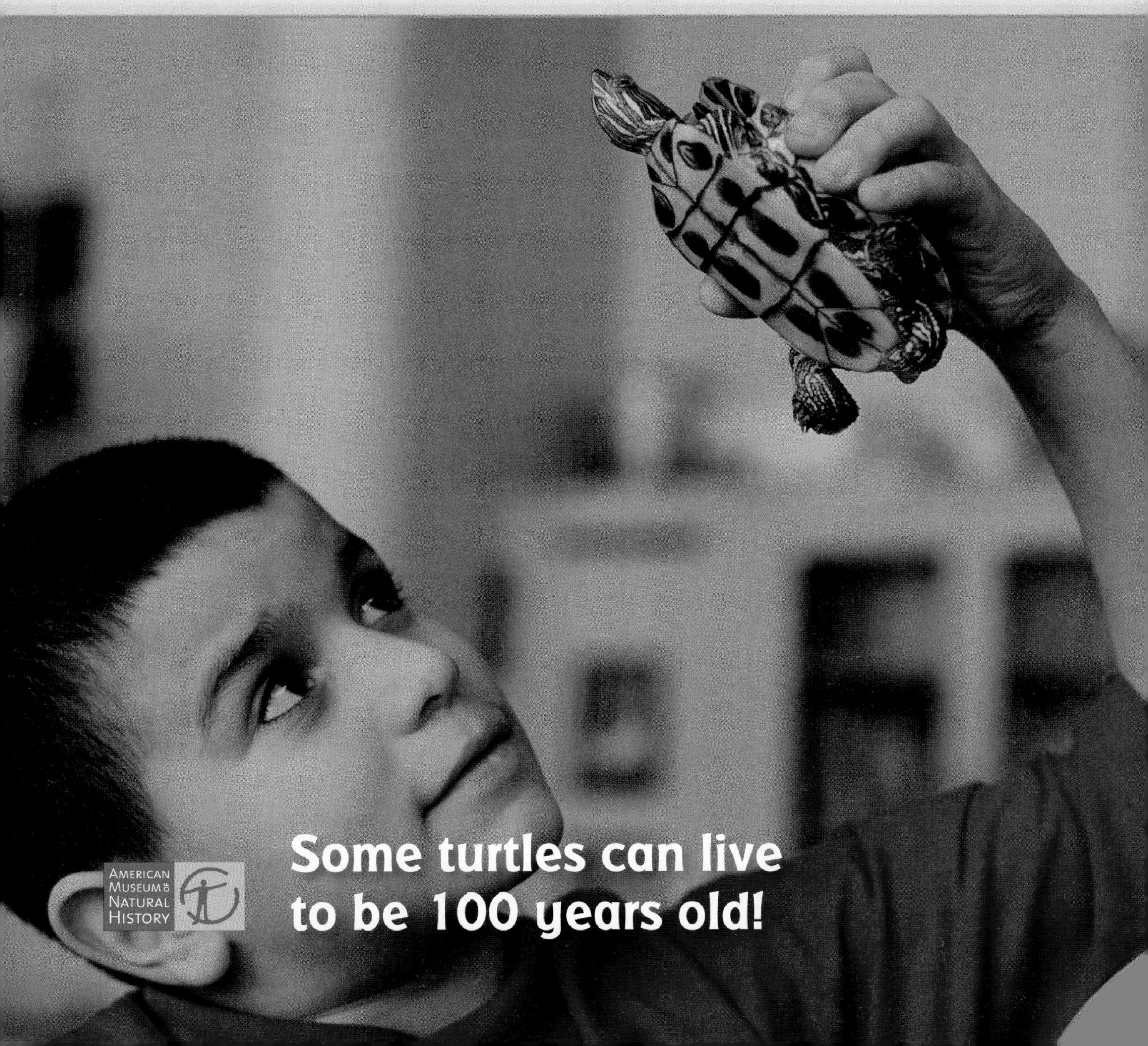

Some turtles can live to be 100 years old!

AMERICAN MUSEUM OF NATURAL HISTORY

Science Skills

Look and Wonder

Have you ever wondered about animals? Scientists wonder about things in our world, too.

Explore

Inquiry Activity

What do you notice about these animals?

What to Do

1. Look at the animals on this page.
2. How are the animals alike? How are they different?
3. Put the animals into groups. Tell a classmate why you put the animals in each group.

Explore More

4. Think of other animals that you could add to your groups.

What do scientists do?

You observed animals to see what they were like. Scientists observe things, too. You can be a scientist!

When you **observe** something, you carefully look, hear, taste, touch, or smell it.

Observe What can you observe about these animals?

dog

fish

bird

butterfly

Scientists can compare and classify animals to learn more about them.

Compare means to see how things are alike or different. **Classify** means to group things by how they are alike.

Compare and Classify

Sort these animals into groups.

koala

flamingo

ladybug

snake

How do scientists work?

Scientists also measure things. **Measure** means to find out the size or amount of something.

Measuring can help scientists **put things in order**, or tell which comes first, next, and last.

butterfly

Measure Use a ruler to measure these insects.

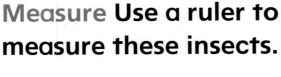

bee

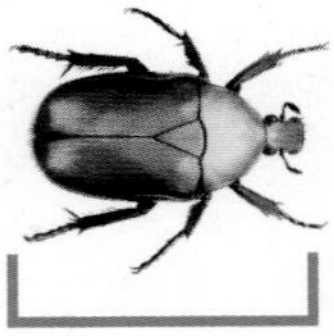

beetle

fly

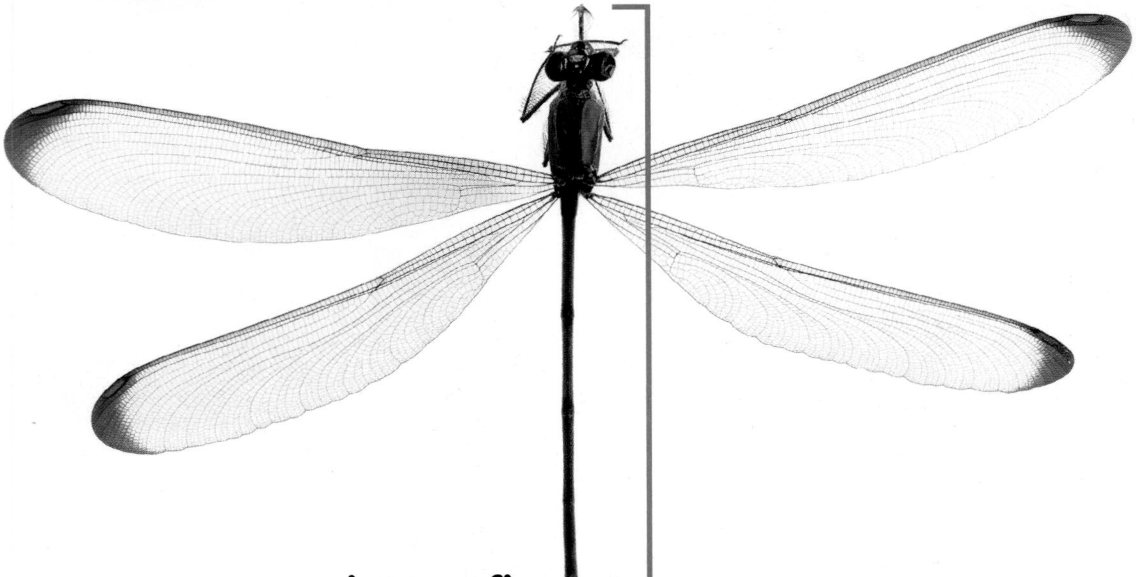

dragonfly

Scientists can make a chart to **record data**, or write down what they observe. They can also **make a model** to show how something looks or works.

When you write or tell your ideas to others you **communicate** what you have learned.

Make a Model
What does this model show?

Communicate
What does this chart tell you about insects?

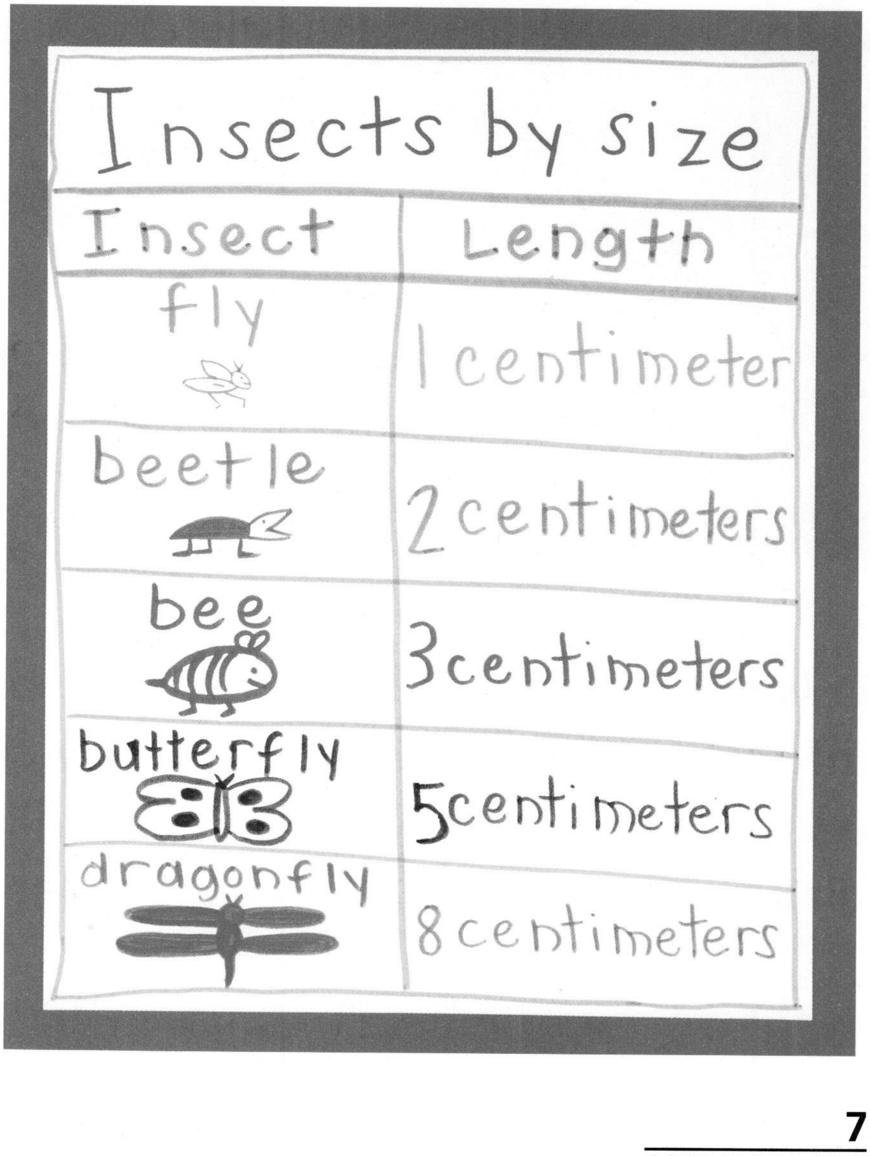

Insects by size

Insect	Length
fly	1 centimeter
beetle	2 centimeters
bee	3 centimeters
butterfly	5 centimeters
dragonfly	8 centimeters

How do scientists learn new things?

Scientists **infer**, or use what they know to figure something out.

Scientists also **predict**, or use what they know to tell what they think will happen.

Infer

It is winter. You can infer that there is not a lot of food for the bird to eat.

Predict

The bird is hungry. You can predict that it will eat the food in the feeder.

Then scientists investigate by making a plan and following it.

They can draw conclusions, or use what they observe to explain what happens.

Investigate

To investigate, you can try to feed the bird different seeds.

Draw Conclusions

If there are no sunflower seeds left, you can draw the conclusion that the bird liked them best.

Think, Talk, and Write

1. Describe what scientists do to learn more about our world.
2. Write a list of things that you want to know more about.

Scientific Method

Look and Wonder

Have you ever wondered about snails? What could you do to find out about them?

Explore

Inquiry Activity

Where can snails live?

What to Do

1. **Observe.** Snails can live in ponds or gardens. Look closely at the pictures of each one.
2. **Compare.** How is the pond the same as the garden? How is it different?
3. **Record Data.** Draw and label the things you see in the pond and the garden.
4. **Draw a Conclusion.** What do you think a garden snail might eat? What might a pond snail eat? Why?

garden

pond

What do snails like to eat?

Scientists make plans. Their plan is called the scientific method. You can use this plan, too!

Mr. Lopez's science class made a plan to find out what snails eat.

Observe

Ask a Question

What do garden snails eat?

Make a Prediction

Garden snails eat garden plants.

Like scientists, the children wrote their plan down so others could follow it.

The plan was to give the snails garden plants and jelly beans. Then the children observed the snails and recorded what they ate.

Make a Plan

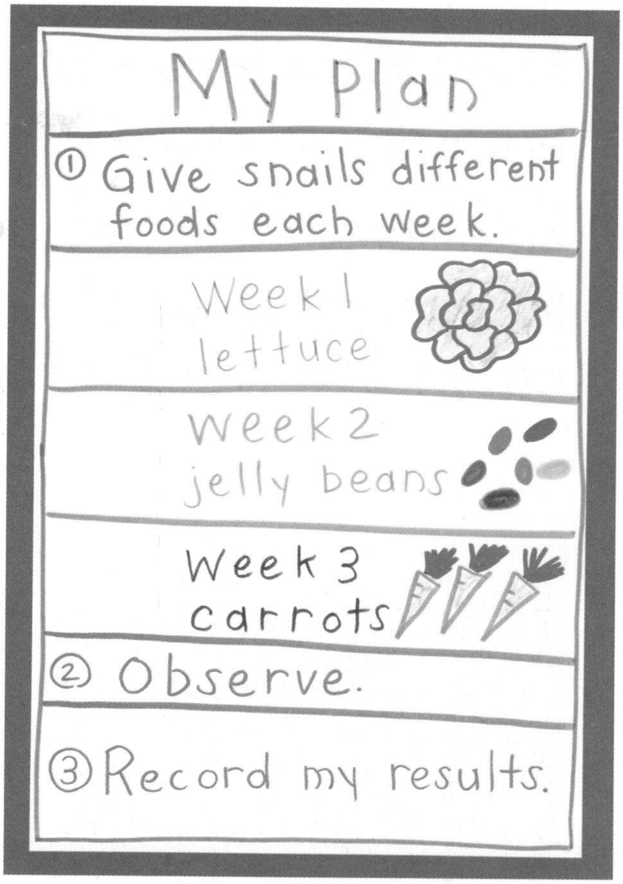

Follow the Plan

Record the Results

What did the children find out?

The children found out that snails eat garden plants.

Like scientists, they wanted to be sure. They tried their experiment again. They got the same results.

Try the Plan Again

Our Results

Food	First try	second try
lettuce	snails ate	snails ate
jelly beans	snails did not eat	snails did not eat
carrots	snails ate	snails ate

Draw a Conclusion

Snails do eat garden plants!

The children shared what they learned about snails with their classmates.

This can lead to new questions and investigations.

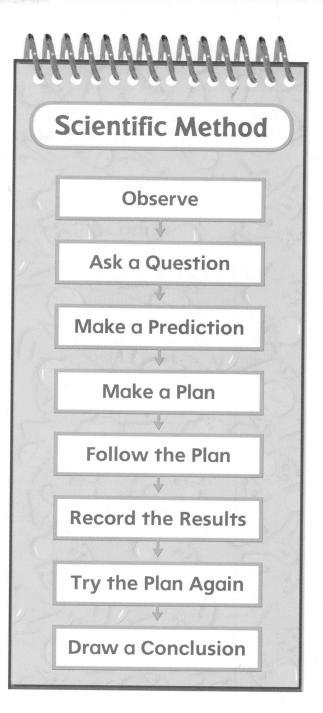

Scientific Method

- Observe
- Ask a Question
- Make a Prediction
- Make a Plan
- Follow the Plan
- Record the Results
- Try the Plan Again
- Draw a Conclusion

Think, Talk, and Write

1. Describe what the children did to find out what snails ate.
2. Write a question you have about snails.

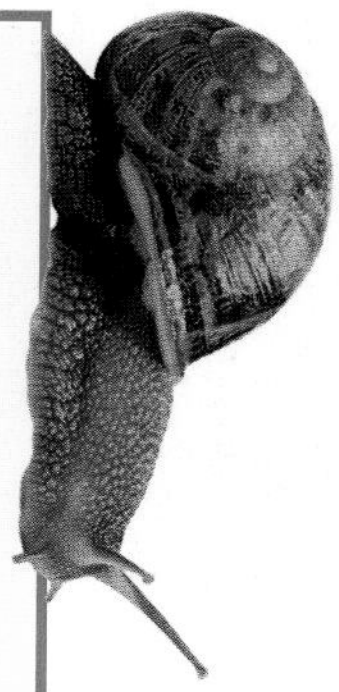

Safety Tips

When you see △ Be Careful, follow the safety rules.

Tell your teacher about accidents and spills right away.

Be careful with sharp objects and glass.

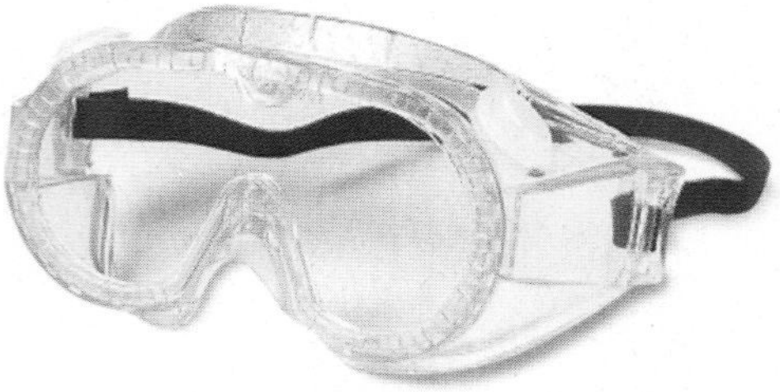

Wear goggles when you are told to.

Wash your hands after each activity.

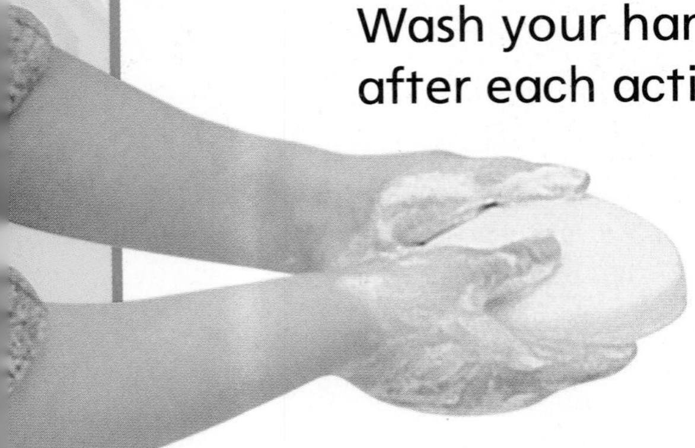

Keep your workplace neat. Clean up when you are done.

UNIT A

Plants

Sunflowers turn to face the Sun all through the day.

Literature

National Wildlife Federation Ranger Rick

Magazine Article

ant stuck on a sundew plant

Sticky! The ant is stuck. The leaf slowly closes and the ant is lunch.

from *Ranger Rick*

Insect-Eating Plants

Did you know that some plants can eat animals?

Venus's-flytrap

Snap! The leaf closes. Long spines hold the grasshopper inside. Juices from the plant digest the insect.

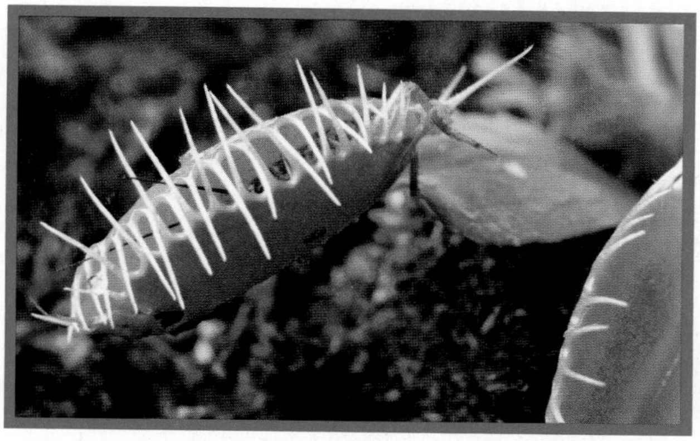

▲ grasshopper caught in a Venus's-flytrap

Pitcher Plant

Splash! Flies fall into the cup shaped leaf. They can not crawl out. The plant eats the insects.

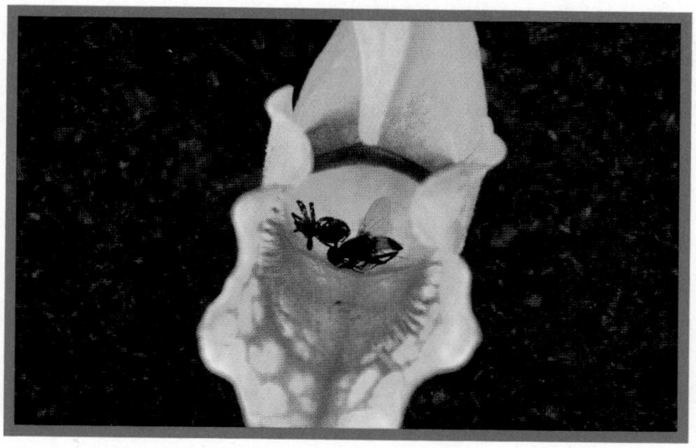

▲ flies trapped in a pitcher plant

Talk About It

How do most plants get what they need?

CHAPTER 1

Plants Are Living Things

The Big Idea

What do you know about plants?

Grand Teton National Park in Wyoming

Key Vocabulary

living something that grows, changes, and needs food, air, and water to survive (page 24)

nonliving something that does not grow and change, or need food, air, or water to survive (page 25)

leaves plant parts that use sunlight and air to make food (page 32)

root plant part that keeps the plant in the ground (page 32)

More Vocabulary

nutrient, page 26

stem, page 32

trunk, page 39

Lesson 1

Learning About Living Things

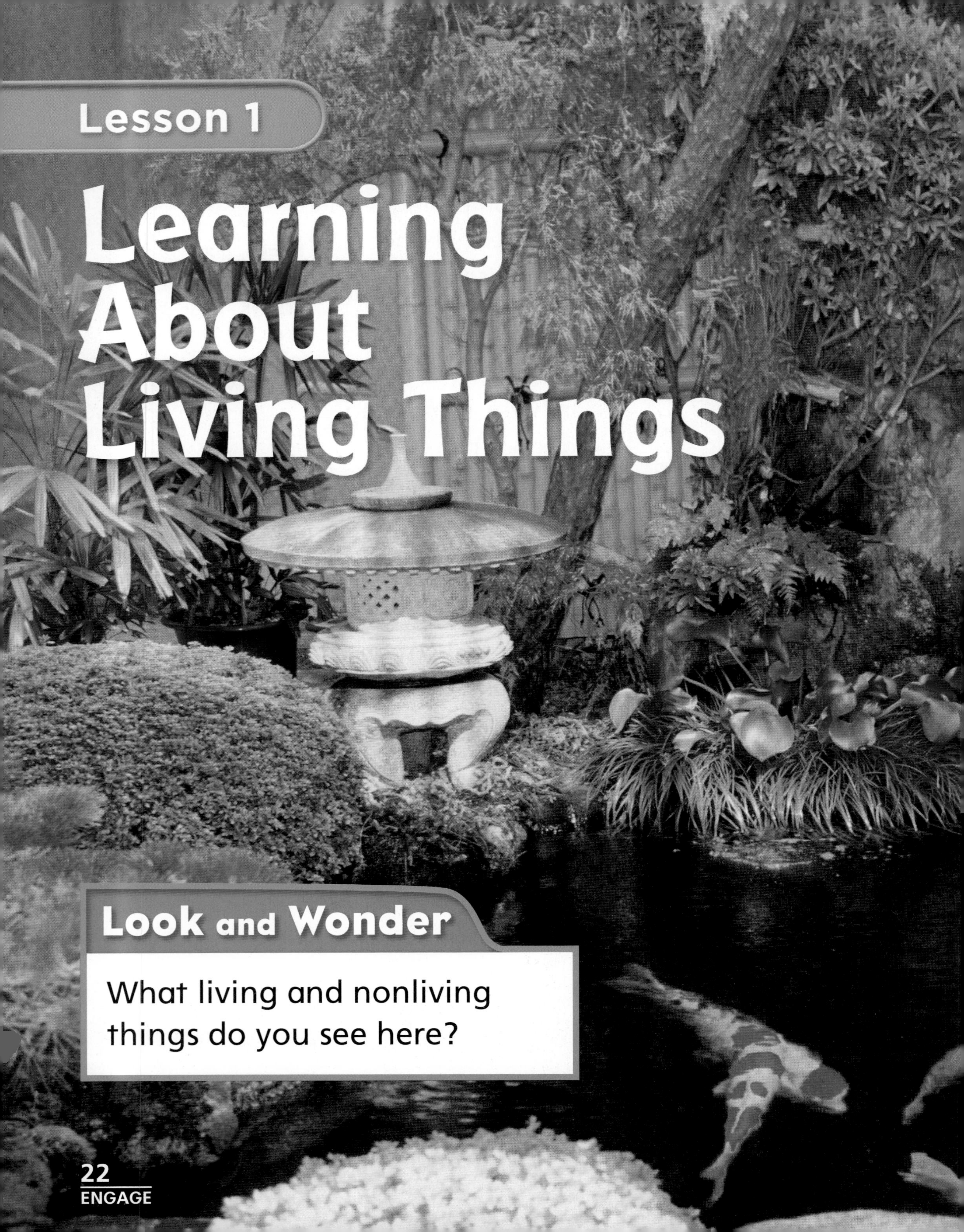

Look and Wonder

What living and nonliving things do you see here?

Explore

Inquiry Activity

What is living and nonliving?

What to Do

1. **Compare.** Look at a rock and a plant. Write about how they are alike and different.
2. Put the rock in a bin. Water the rock and the plant for a week.
3. **Observe.** What happens?
4. **Infer.** How do you know if something is living or nonliving?

Explore More

5. **Classify.** Sort living and nonliving objects.

You need

rock

plant

water

clear bin

Read Together and Learn

Vocabulary

living

nonliving

nutrient

SCIENCE QUEST Explore the needs of plants with the Junior Rangers.

What are living and nonliving things?

You are a living thing. Plants and animals are, too. **Living** things grow and change.

Living things need food, water, and air to live. They make new living things like themselves.

Living and Nonliving Things

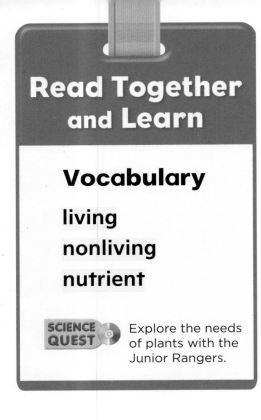

Nonliving things do not grow and change.

They do not need food, water, or air to survive. They do not make new things like themselves.

Quick Lab

Find living and nonliving things in your classroom.

✓ **How are living and nonliving things different?**

Read a Photo

What living and nonliving things do you see here?

Why are plants living things?

Plants need air, water, nutrients, sunlight, and space to live and grow. Just like food helps you grow, **nutrients** help plants grow.

Plants grow where they get what they need to survive.

▲ **Rice plants need a lot of water to survive.**

▲ **Cactuses do not need a lot of water to survive.**

FACT A cactus plant can hold enough water inside it to last a whole year.

Plants use water, air, and sunlight to make their own food.

Plants grow and change. They make new plants like themselves.

Sunflowers need a lot of sunlight to live.

✔ **How are you different from plants?**

Think, Talk, and Write

1. **Main Idea and Details.** Is a car living or nonliving? Why?

2. Write about what plants need to live.

Art Link

Draw a picture with living and nonliving things in it. Label your picture.

LOG ON e-Review Summaries and quizzes online at www.macmillanmh.com

Lesson 2

Parts of Plants

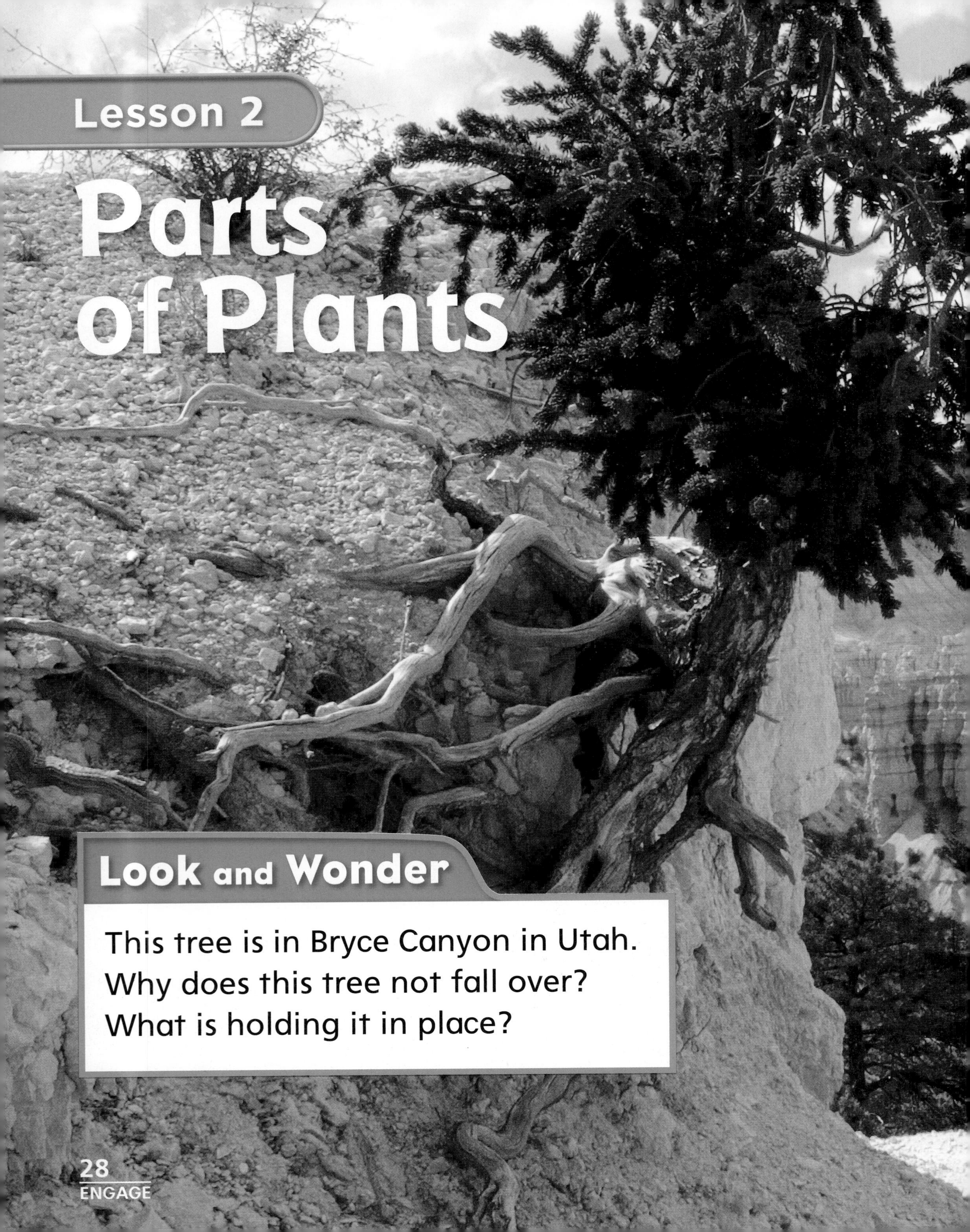

Look and Wonder

This tree is in Bryce Canyon in Utah. Why does this tree not fall over? What is holding it in place?

Explore

Inquiry Activity

What are the parts of a plant?

You need

plant

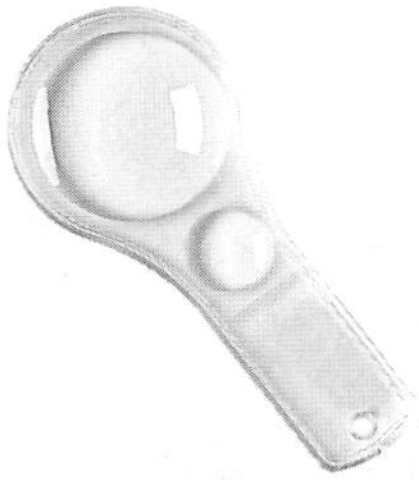

hand lens

What to Do

1. Gently pull a plant from the soil.
2. **Observe.** Use a hand lens to look at the plant's stems, leaves, and roots.
3. **Communicate.** Draw a picture of the plant and its parts.

Explore More

4. **Infer.** Why do you think the plant has different parts?

Step 2

Read Together and Learn

Vocabulary

leaves

stem

root

SCIENCE QUEST Explore leaves, stems, and roots with the Junior Rangers.

What are the parts of plants?

Plants can not walk around like you. They have to get everything they need from the space around them.

Plants have parts to help them get what they need. Most plants have leaves, stems, and roots.

These parts can look different on different kinds of plants.

What are some parts of a plant?

Strawberry Plant

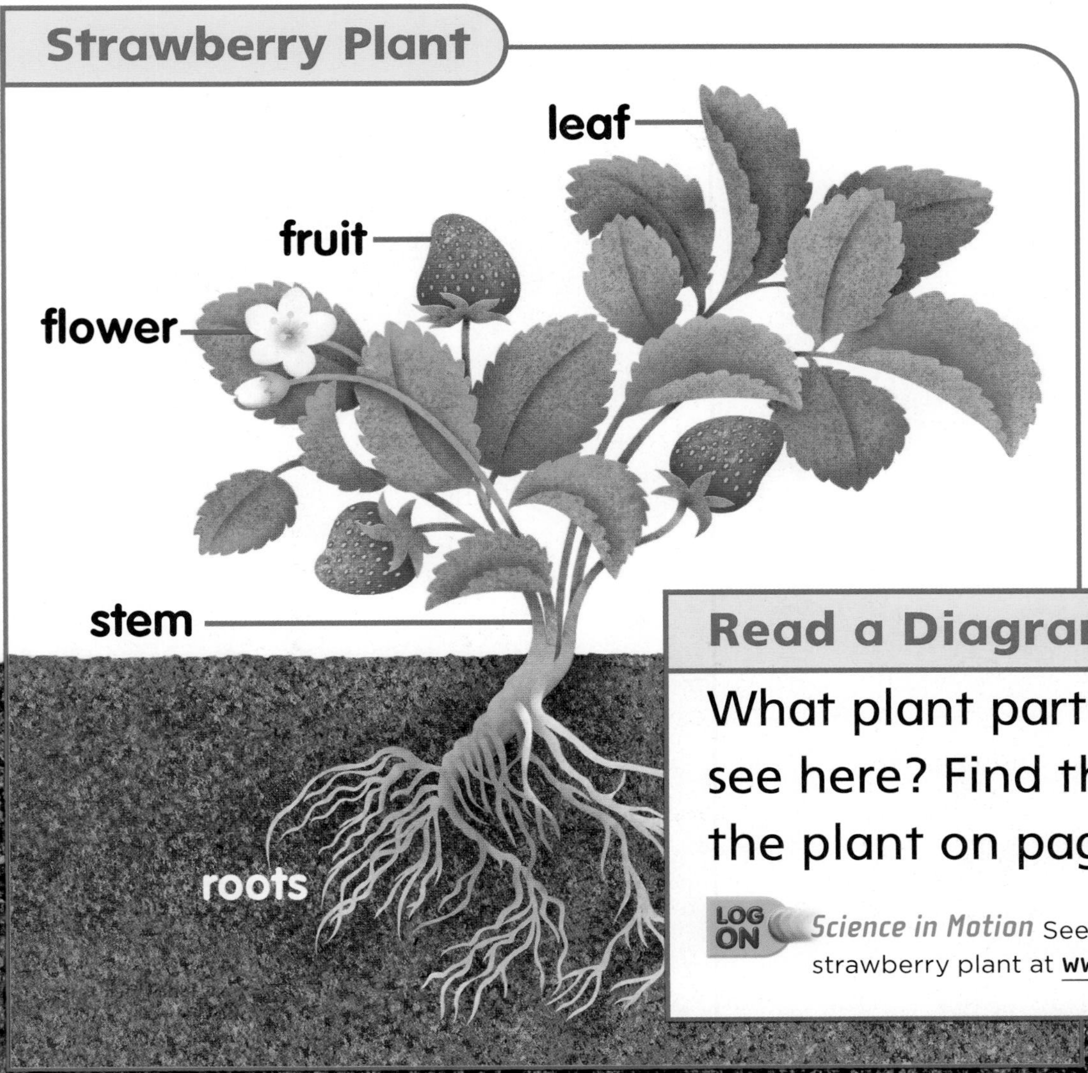

Read a Diagram

What plant parts do you see here? Find them on the plant on page 30.

LOG ON *Science in Motion* See the parts of a strawberry plant at **www.macmillanmh.com**

What do plant parts do?

Leaves use sunlight and air to make food. Food and water move through the stem to other plant parts.

The **stem** holds up the plant. **Roots** keep the plant in the ground.

Quick Lab

Put celery in colored water. Draw what you see.

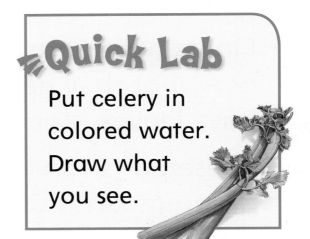

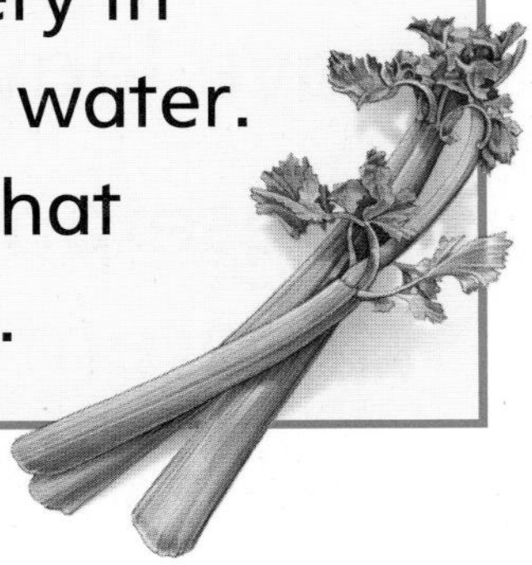

Plants have roots that take in water and nutrients from the soil.

Some plants have roots close to the surface of the ground. Others have long and deep roots.

roots close to the surface of the ground

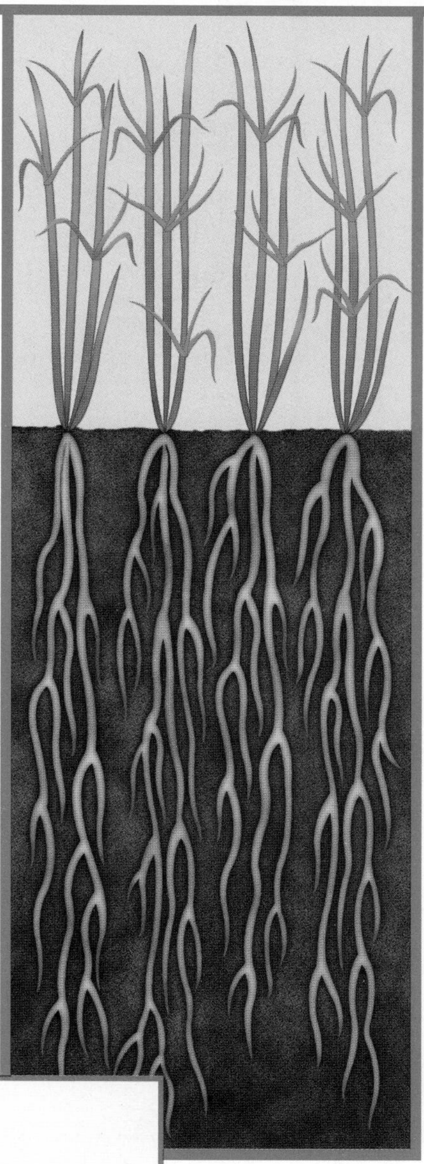

long and deep roots

✓ Why are roots important?

Think, Talk, and Write

1. **Summarize.** Describe how each plant part helps a plant survive.
2. Write about why plants have leaves.

Math Link

Find two plants. Measure the length of their stems. Compare their sizes.

LOG ON e-Review Summaries and quizzes online at www.macmillanmh.com

Meet General Sherman

A tree named General Sherman in California's Sequoia National Park is said to be the largest in the world.

It is 275 feet tall. That is as tall as a building with 27 floors!

Write About It

Write about a tall plant that you have seen. Draw it and label its parts.

Remember
Use words that describe the plant.

LOG ON e-Journal Write about it online at www.macmillanmh.com

Seeds of All Sorts

Michael sorted his seeds. He made a picture graph to show how many of each seed he has.

Michael's Seeds	
beans	
sunflower seeds	
corn kernels	
peas	

Read a Graph

Does Michael have more sunflower seeds or beans? Write a number sentence to show how you know.

If he found 6 more peas, how many would he have? Write a number sentence to show the answer.

Remember
A picture graph helps you solve a problem.

Lesson 3

Different Plants

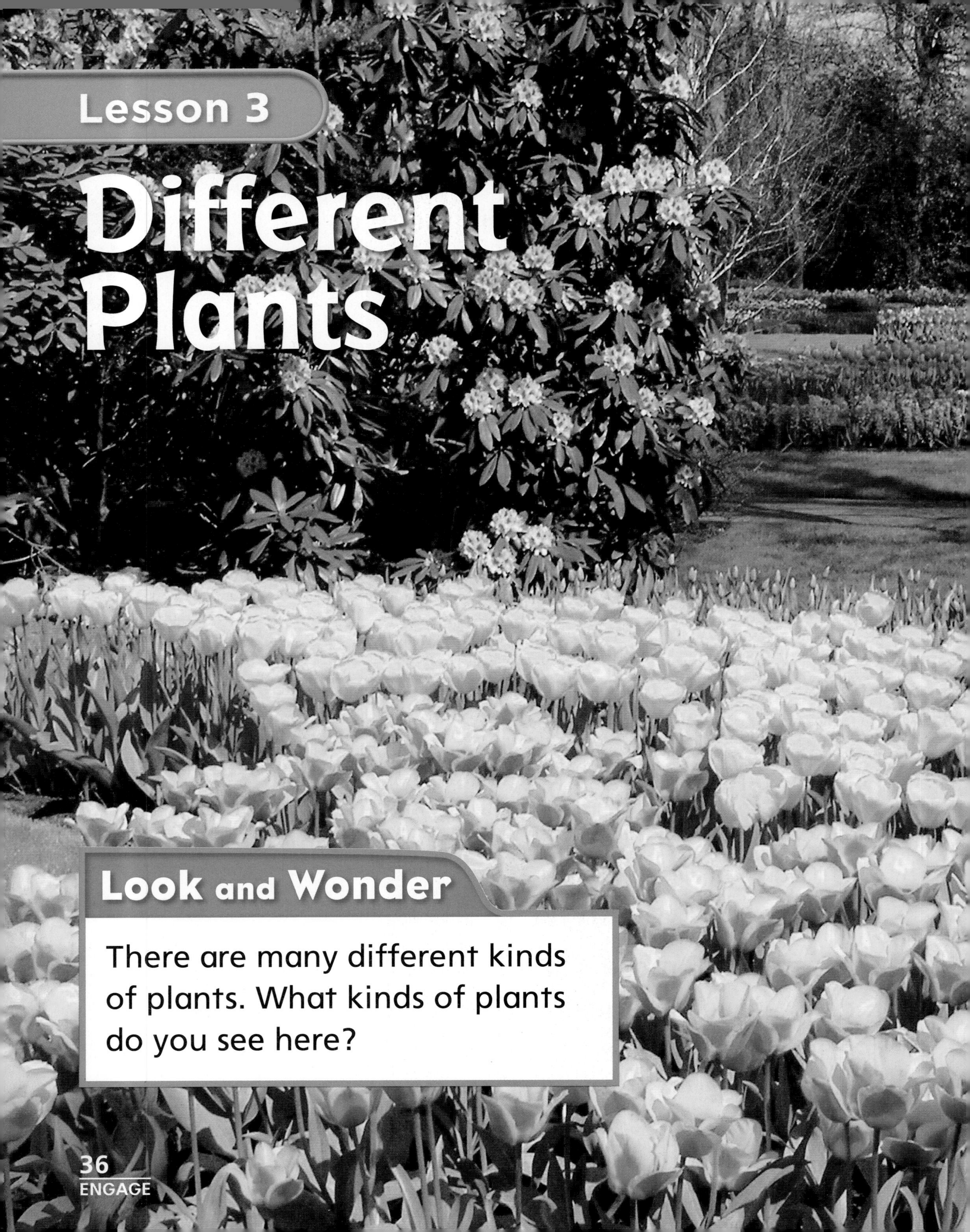

Look and Wonder

There are many different kinds of plants. What kinds of plants do you see here?

Explore

Inquiry Activity

How are plants different?

You need

paper

crayons

What to Do

1. **Observe.** Find some plants around your school. How many different plants can you find?
2. **Record Data.** Draw two different plants that you find.
3. **Compare.** How are the plants alike? How are they different?

Explore More

4. **Communicate.** How could you find out more about the plants you saw?

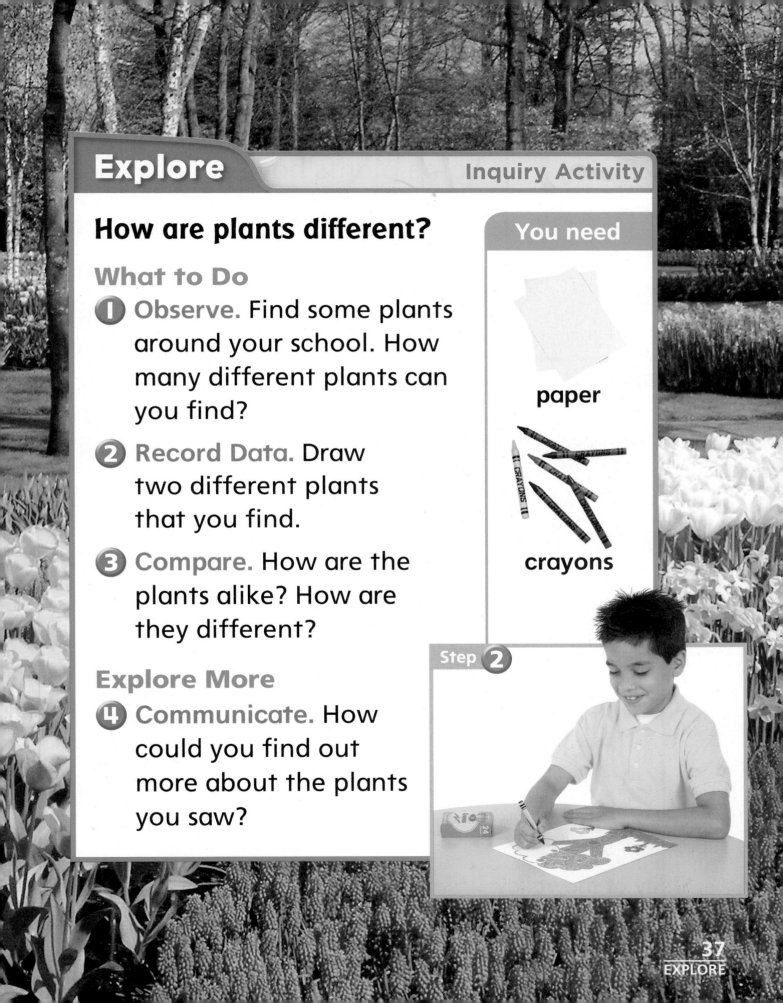

Step 2

Read Together and Learn

Vocabulary

trunk

How are plants different?

Not all plants look the same. Some plants have colorful flowers. Other plants are only green.

Different plants can have different types of leaves.

Stems can be thin or thick. The thick stem of a tree is called a **trunk**.

Some plants are tall. Other plants, like grass, spread out along the ground.

✔ **How are these plants alike? How are they different?**

Tall Trees

Read a Photo

How do you know the trees are tall?

grass

Which plant parts can you eat?

You can eat different plant parts. Some plant parts are safe to eat.

Others are not. Eating parts of some plants can make you sick.

Quick Lab

Find out which plant parts your classmates ate today.

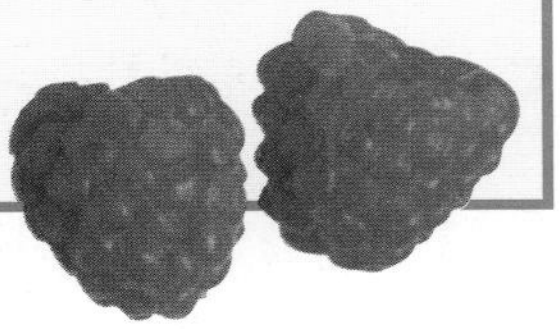

▲ **When you eat a coconut, you eat the seed of a plant.**

▲ **When you eat cinnamon, you eat a part of a tree trunk.**

When you eat lettuce, you eat the leaves of a plant. ▼

When you eat a carrot, you eat the root of a plant. ▼

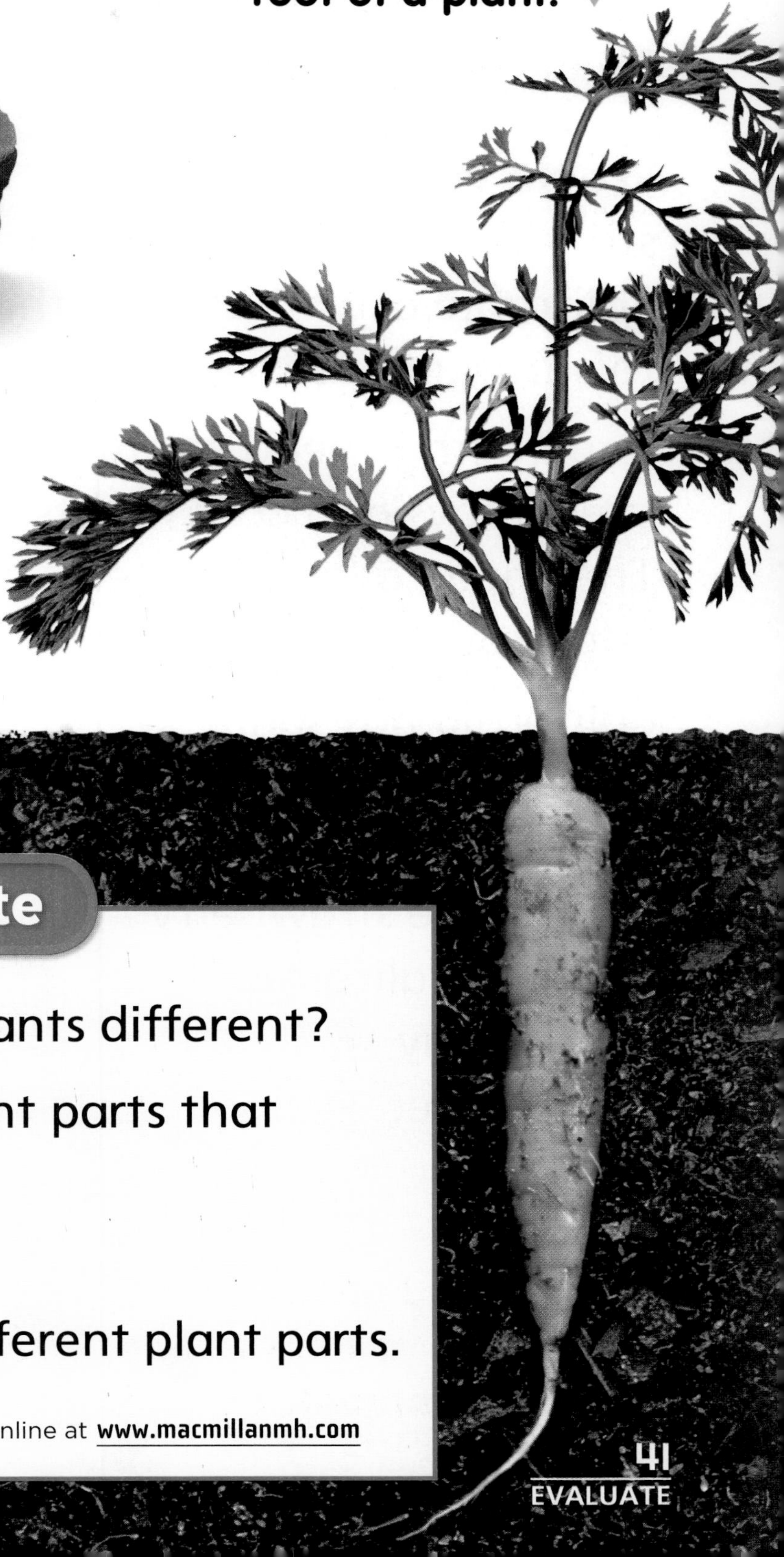

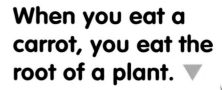

✔ What plant parts do you eat?

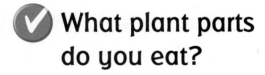

Think, Talk, and Write

1. Classify. How are plants different?
2. Write about the plant parts that you ate today.

Art Link

Make a collage with different plant parts.

LOG ON e-Review Summaries and quizzes online at www.macmillanmh.com

Strawberry Fields

Would you like to grow strawberries? The Southern California coast has everything this fruit needs to grow.

Strawberries need warm, sunny days and cool nights. They need a lot of water and grow best in sandy soil.

California has sandy soil and is sunny all year. That is why people plant more strawberry farms in California than in any other state.

American Museum of Natural History

Talk About It

Main Idea and Details. What do strawberry plants need to grow?

I Read to Review

My Plant Book

Some roots are thick.

Some roots are thin.

Some stems are thick.
Some stems are thin.

Some leaves are little.

Some leaves are big.

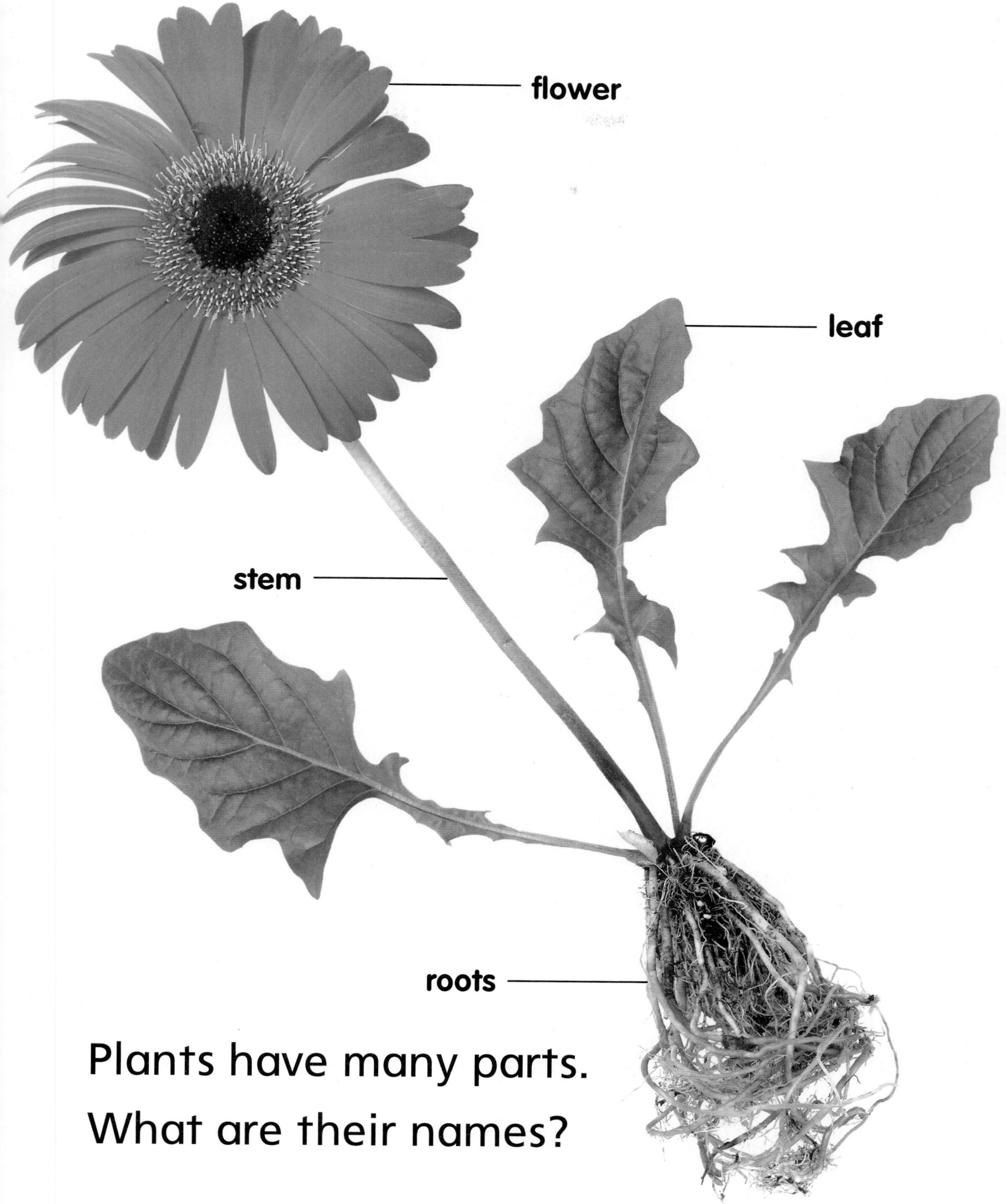

Plants have many parts.
What are their names?

CHAPTER 1 Review

Vocabulary

Use each word once to complete the sentences.

leaves
roots
stems

1. These are both _____.

2. These are both _____.

3. These are both _____.

Answer the questions below.

4. Which is living and which is nonliving? Tell how you know.

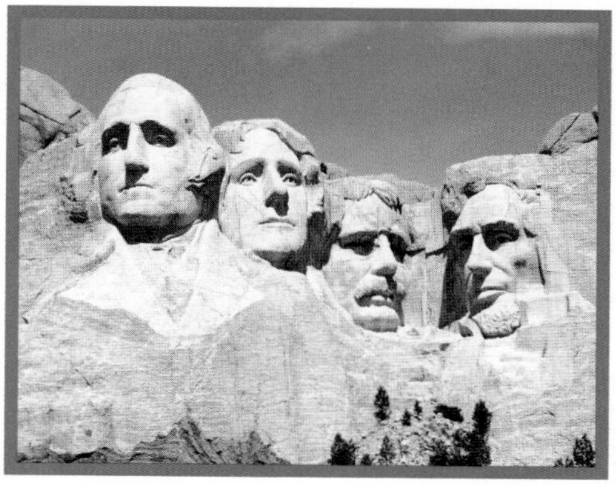

5. **Observe.** How are these plants getting what they need to live?

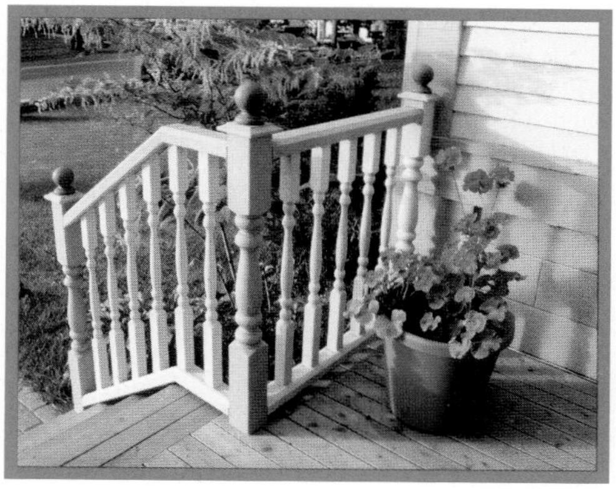

6. **Main Idea and Details.** Name the parts of plants and what they do.

The Big Idea

7. What do you know about plants?

CHAPTER 2

Plants Grow and Change

The Big Idea

How do plants change?

Key Vocabulary

flower a part of a plant that makes seeds (page 54)

seedling a young plant (page 60)

desert a hot and dry place (page 68)

arctic an icy and cold place near the North Pole (page 70)

More Vocabulary

seed, page 54

fruit, page 55

life cycle, page 60

rain forest, page 69

Lesson 1

Flowers, Fruits, and Seeds

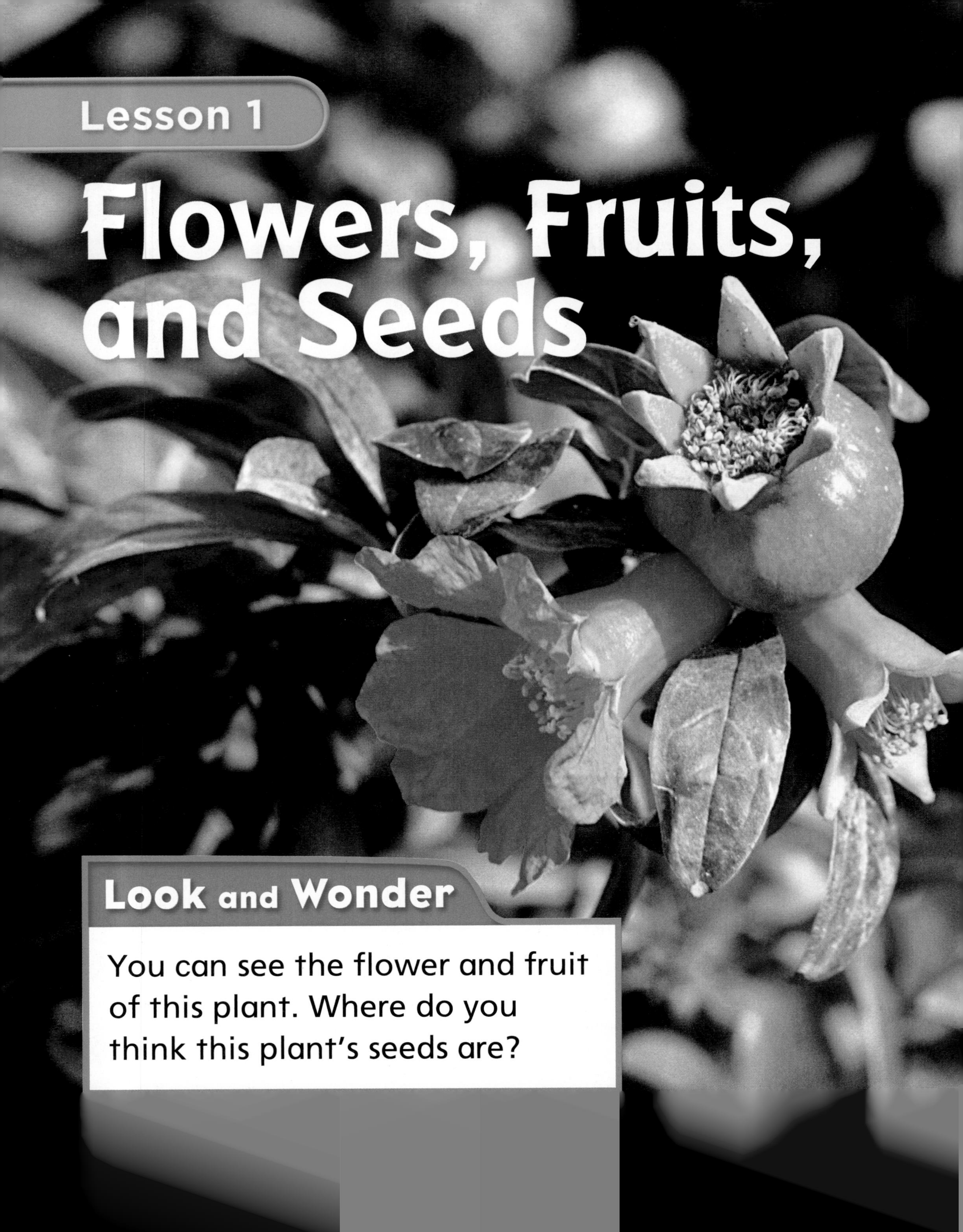

Look and Wonder

You can see the flower and fruit of this plant. Where do you think this plant's seeds are?

Explore

Inquiry Activity

How can you classify seeds?

You need

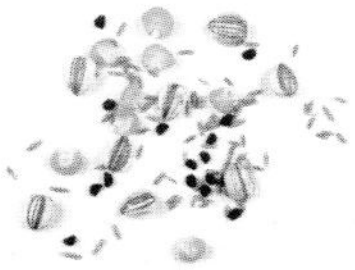

seeds

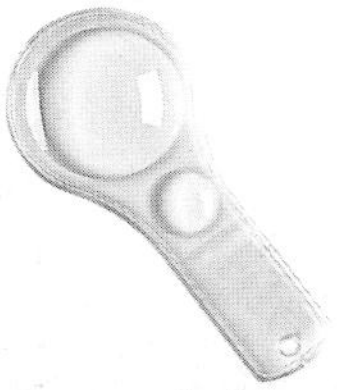

hand lens

glue

What to Do

1. **Observe.** Look at seeds with a hand lens.
2. **Classify.** Sort the seeds into groups. How did you sort the seeds?
3. **Record Data.** Make a chart to show how you sorted the seeds. Glue your seed groups onto the chart.

Explore More

4. **Compare.** Which group has the most seeds? Which group has the fewest seeds?

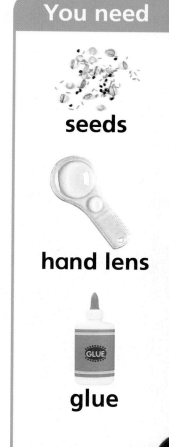

Step 3

Read Together and Learn

Vocabulary

flower

seed

fruit

SCIENCE QUEST Explore flowers, fruits, and seeds with the Junior Rangers.

Why are flowers important?

Living things are made of parts. Plants have different parts. Some plants have colorful flowers.

A **flower** is a part of a plant that makes seeds. A **seed** is a part of a plant that can grow into a new plant.

peach trees

FACT Nectarines are peaches without the fuzz.

Seeds are protected by the fruit of some plants. A **fruit** is the plant part that grows around seeds.

Fruits can be juicy and good to eat. We eat the fruits of many plants. Sometimes we eat seeds, too.

▼ **Flowers on peach trees will grow into fruit.**

▼ **The flowers are gone and peaches fill the tree.**

◀ **The fruit protects the seed inside.**

✓ What other fruits do you eat?

What are the parts of a seed?

Seeds come in many different shapes and sizes.

Seeds need water, light, and a warm place to grow. Seeds have parts to help them grow.

Quick Lab

Open a lima bean. Draw and label its parts.

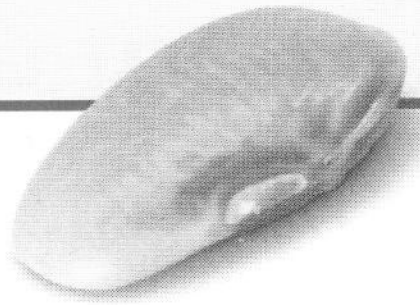

Look Inside a Seed

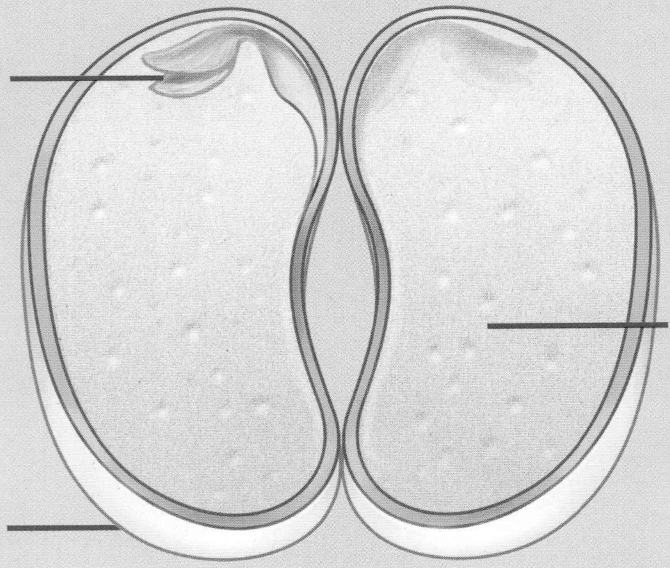
This is a tiny plant inside a seed that will grow into a new plant.
This is food for the tiny plant.
Many seeds have an outer covering that keeps the seed safe.

Read a Diagram

What are the parts of this bean seed?

Seeds do not always grow in the same place. Wind can move seeds to a new place.

Animals can also move seeds. These seeds will grow plants in new places.

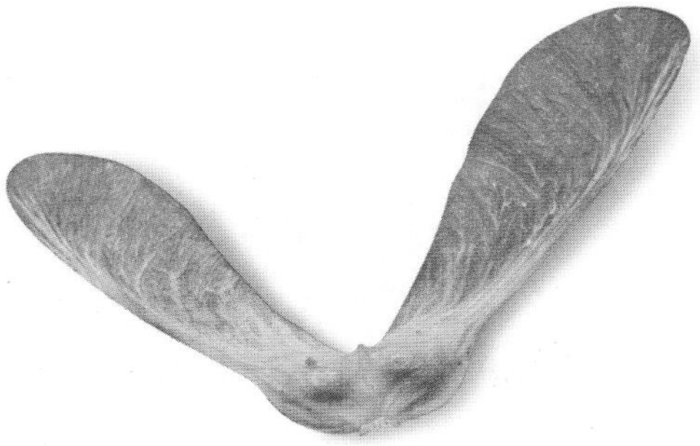

▲ **Wind can carry maple seeds to new places to grow.**

✓ **What is inside a seed?**

▶ **Squirrels bury seeds. If they do not go back to eat the seeds, a new plant will grow there.**

Think, Talk, and Write

1. **Summarize.** Why are flowers and fruits important to some plants?

2. Write about how seeds can move to different places.

Art Link

Use different seeds to make a collage.

LOG ON e-Review Summaries and quizzes online at www.macmillanmh.com

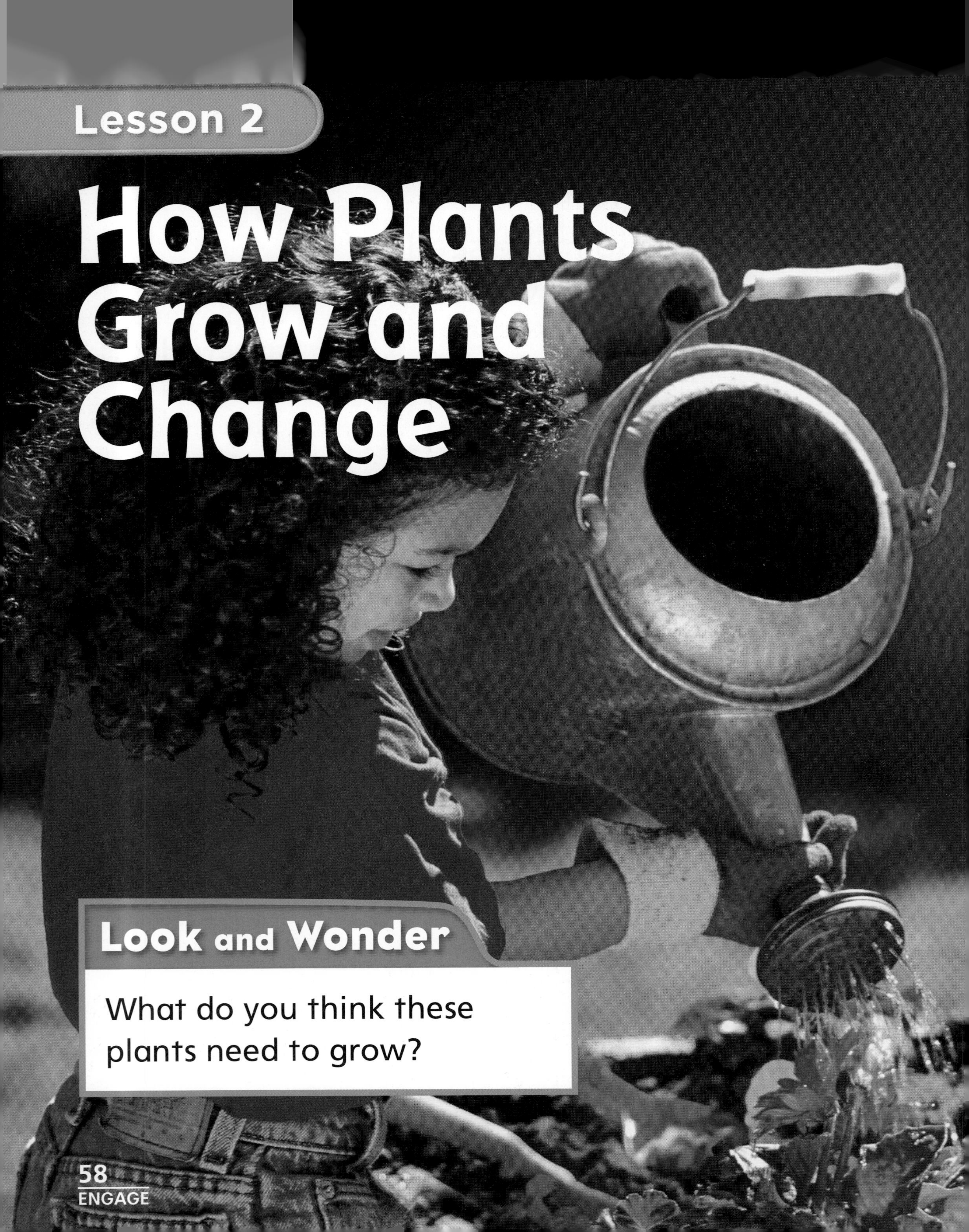

Lesson 2

How Plants Grow and Change

Look and Wonder

What do you think these plants need to grow?

Explore

Inquiry Activity

What do seeds need to grow?

What to Do

1. Put some seeds on a wet paper towel. Put some on a dry towel. Put each paper towel in a bag.
2. **Classify.** Label the bags "Dry" and "Wet." Place both bags in a warm place.
3. **Record Data.** Look at your seeds. Draw and write about what happens to them.

Explore More

4. **Investigate.** What will happen if you water both seeds but only one seed has light?

You need

seeds

paper towels

water

ziplock bags

Step 3

Read Together and Learn

Vocabulary

life cycle

seedling

How do plants grow from seeds?

Some plants' life cycles begin with a seed. A **life cycle** shows how a living thing grows, lives, and dies.

A seed grows when it gets water and nutrients from soil. The seed sprouts and becomes a young plant that is called a **seedling**.

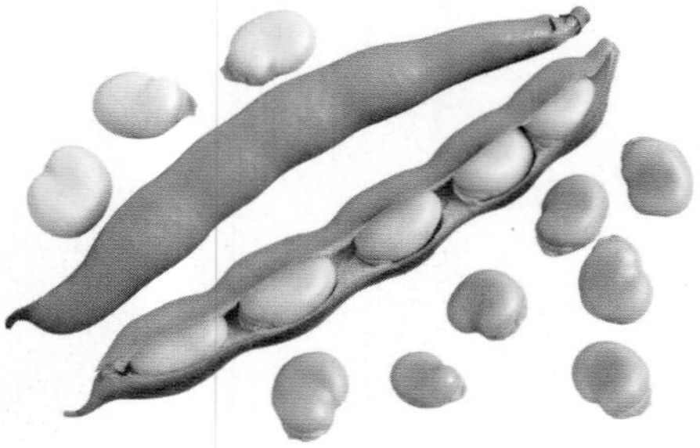

bean seeds

Life Cycle of a Bean Plant

FACT Beans are fruits. Fact_Txt

Soon the seedling grows and becomes an adult plant. It looks like the plant it came from.

The new plant will make more seeds and the life cycle goes on.

✔ What is the life cycle

Read a Diagram

How do bean plants grow and change?

LOG ON *Science in Motion* Watch the life cycle of a bean plant at **www.macmillanmh.com**

How else do plants grow?

Plants do not always grow from seeds.

Sometimes when you cut a part of a plant, a new plant will grow from that plant part.

▲ **If you cut a coleus leaf and put it in water, it will grow roots.**

coleus plant

Sometimes you can even grow a new plant from a whole parent plant.

Quick Lab

Grow a new potato plant from a potato.

✔ **What are different ways that plants can grow?**

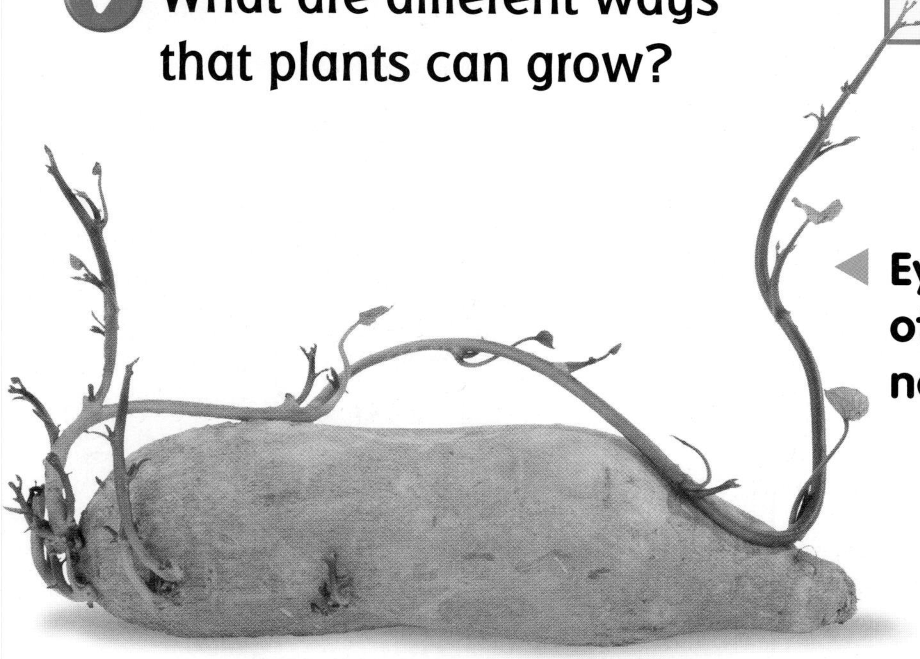

Eyes on the outside of a potato can sprout new potato plants.

Think, Talk, and Write

1. **Compare and Contrast.** How can different plants grow and change?
2. Write about the life cycle of a bean plant.

Social Studies Link

Describe how plants grow and change where you live.

LOG ON e-Review Summaries and quizzes online at www.macmillanmh.com

A GIANT GRASS

Did you know that bamboo is a kind of grass? It is very strong. Bamboo can be used to make many things.

Sometimes people use bamboo to make floors and roofs for their homes. Bamboo can even be used to make sweet treats and drinks.

People can climb on bamboo to fix buildings.

18 feet

25 feet

Some bamboo plants can grow taller than a giraffe!

Talk About It

Summarize. What is bamboo? What can it be used for?

AMERICAN MUSEUM OF NATURAL HISTORY

Lesson 3

Plants Live in Many Places

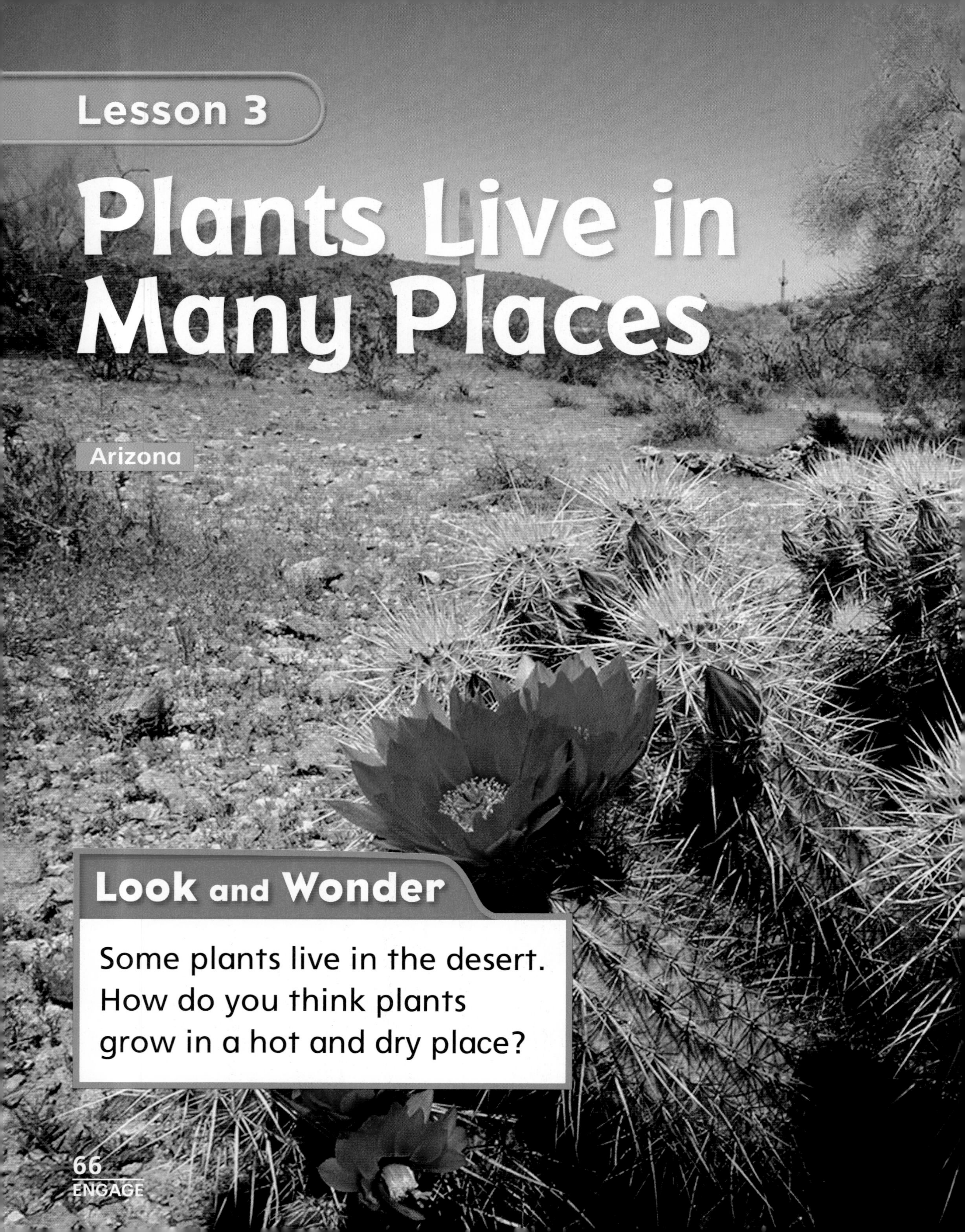

Look and Wonder

Some plants live in the desert. How do you think plants grow in a hot and dry place?

Explore

Inquiry Activity

What happens if a plant does not get water?

What to Do

1. Put two plants in a sunny place. Water only one of the plants.
2. **Predict.** What will happen to each plant?
3. **Observe.** Watch your plants for a week.
4. **Record Data.** Draw what happens to the plants.

Explore More

5. **Classify.** What kinds of plants live in dry or wet places?

You need

two labeled plants

water

Step 1

Read Together and Learn

Vocabulary

desert

rain forest

arctic

Where do plants live?

Plants can live almost anywhere on Earth. They grow where they get what they need to live.

Some plants live in the desert. The **desert** is a dry place. Many deserts are hot. Plants have parts that help them store water in the desert.

▼ **Many desert plants have thick stems and spiny leaves that help them store water.**

Other plants live in the rain forest. The **rain forest** is a hot, wet place.

Many plants have large, pointy leaves. Extra water falls off these leaves so that plants do not get too much water.

Why can a rain forest plant not live in a desert?

Rain Forest Plants

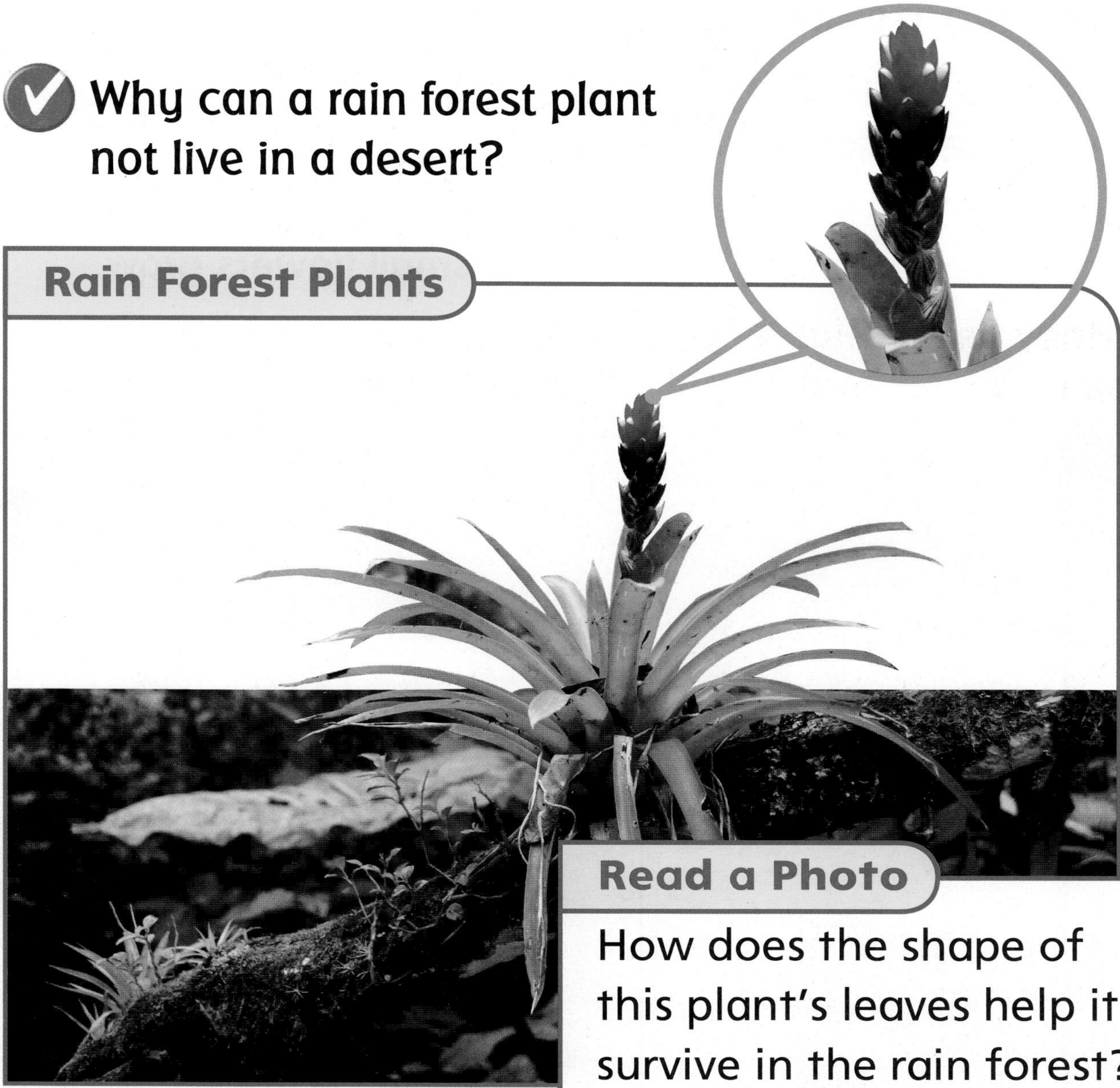

Read a Photo

How does the shape of this plant's leaves help it survive in the rain forest?

How can plants survive in the cold?

Some plants live in the arctic. The **arctic** is an icy and cold place near the North Pole.

Arctic plants grow in groups close to the ground. This keeps them safe from the cold and the wind.

Quick Lab

Put a plant in a freezer for a day. What happens?

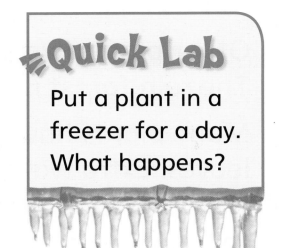

When arctic snow melts, small flowers grow.

Arctic plants also have roots that grow close to the surface of the ground. This helps plants survive in frozen soil.

arctic plants

What helps plants survive in the Arctic?

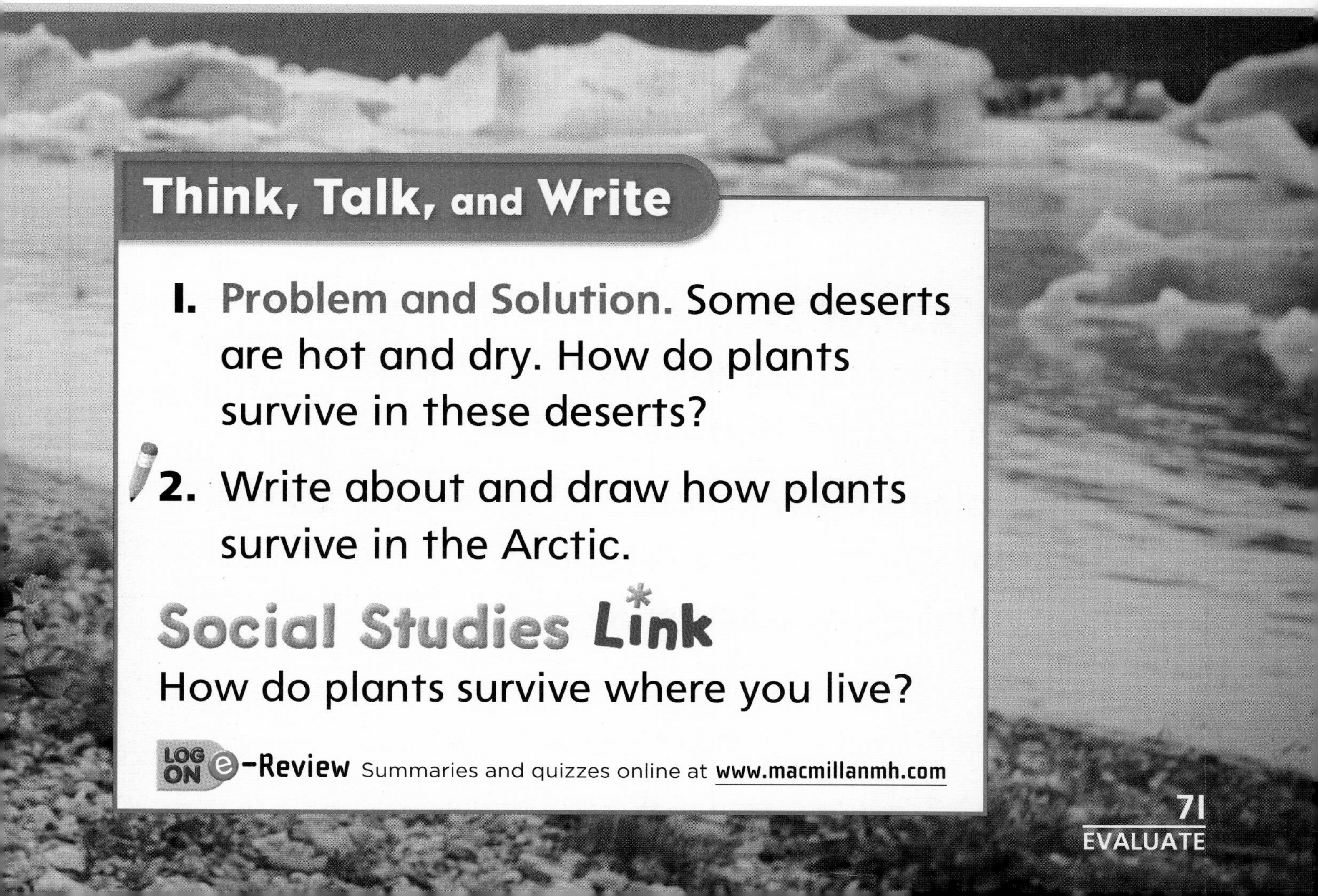

Think, Talk, and Write

1. **Problem and Solution.** Some deserts are hot and dry. How do plants survive in these deserts?
2. Write about and draw how plants survive in the Arctic.

Social Studies Link

How do plants survive where you live?

LOG ON e-Review Summaries and quizzes online at www.macmillanmh.com

Your Own Garden

Some gardens are big. Some are small. People can grow food to eat in their gardens.

Write About It

Write about a garden that you would like to have.

What plants grow in your garden? Are there plants that you could eat in your garden?

Remember

Use words that tell what your garden would look like.

LOG ON e-Journal Write about it online at www.macmillanmh.com

Fruit for Sale

Jasmine wanted to buy some fruit at the market. She had these coins.

Count the Coins

Can Jasmine buy an apple?

Can Jasmine buy a lemon and a lime?

How do you know?

Remember

A quarter is worth 25 cents. A dime is worth 10 cents.

I Read to Review

All About Plants

Some flowers are red.
Some flowers are yellow.

Some seeds are big.
Some seeds are small.

Some plants have fruits.
Some plants do not.

Plants live in different places.
Plants live and grow.

CHAPTER 2 Review

Vocabulary

fruits
life cycle
seedling
seeds

Use each word once to complete the sentences.

1. How a living thing grows, lives, and dies is its ______.

2. These are both ______.

3. A young plant is called a ______.

4. These are both ______.

Science Skills and Ideas

Answer the questions below.

5. How are desert plants like arctic plants? How are they different?

6. **Classify.** Sort the objects below into different groups.

7. **Summarize.** Why are flowers important to some plants?

8. How can seeds move to a new place?

The Big Idea

9. How do plants change?

Park Ranger

Do you want to work with plants? You could become a park ranger.

Some park rangers watch for forest fires. Some care for plants and animals living in a park. Some teach about the park and how to care for it.

park ranger

More Careers to Think About

botanist

marine biologist

LOG ON e-Careers at www.macmillanmh.com

Animals and Their Homes

Dormice sleep all day and all night through the winter.

Literature
Poem

Giraffes

by Mary Ann Hoberman

Giraffes
I like them.
Ask me why.
Because they hold their heads up high.
Because their necks stretch to the sky.
Because they're quiet, calm, and shy.
Because they run so fast they fly.
Because their eyes are velvet brown.
Because their coats are spotted tan.
Because they eat the tops of trees.
Because their legs have knobby knees.
Because
Because
Because. That's why
I like giraffes.

Talk About It

Which words did the poet use to describe giraffes?

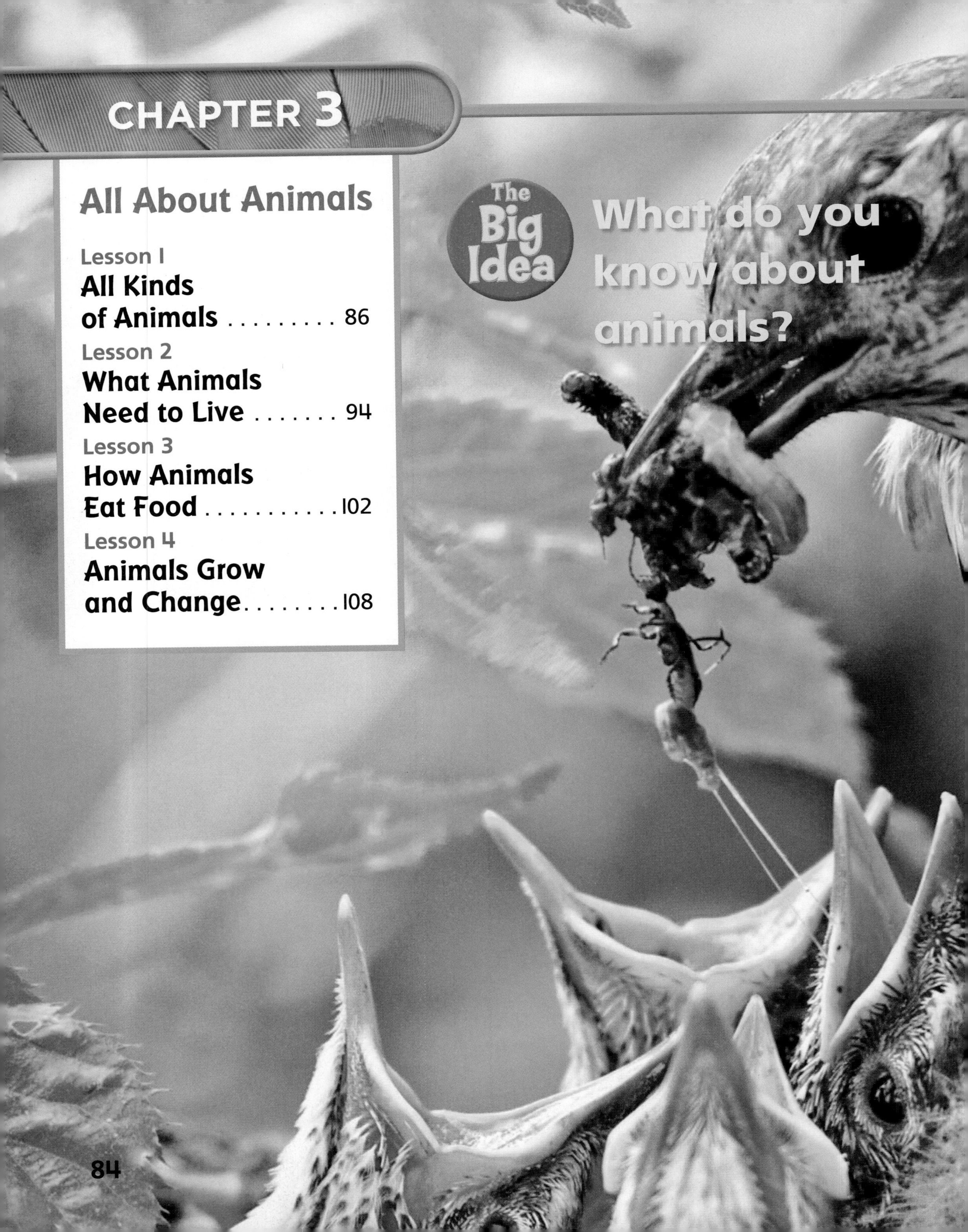

CHAPTER 3

All About Animals

The Big Idea

What do you know about animals?

Key Vocabulary

mammal an animal with hair or fur (page 88)

bird an animal that has two legs, two wings, and feathers (page 89)

reptile an animal that has dry skin covered with scales (page 90)

amphibian an animal that lives on land and in water (page 91)

More Vocabulary

fish, page 92

insect, page 93

shelter, page 97

gills, page 99

lungs, page 99

herbivore, page 105

carnivore, page 106

hatch, page 112

tadpole, page 114

Lesson 1

All Kinds of Animals

Look and Wonder

Are all animals like these puppies? Why or why not?

Explore

Inquiry Activity

What are some different kinds of animals?

What to Do

1. Cut out pictures of different animals.

⚠ **Be Careful.** Remember, scissors can be sharp!

2. **Classify.** Sort the pictures of animals into groups.

Explore More

3. **Compare.** Are your groups the same or different from those of your classmates? What other animals could you put in each group? Why?

You need

magazines

scissors

Read Together and Learn

Vocabulary

mammal
bird
reptile
amphibian
fish
insect

What are some types of animals?

There are many different kinds of animals. **Mammals** are a group of animals with hair or fur. They can hop, walk, swim, or fly.

Porcupines are mammals.

Mammals, like giraffes, take care of and give birth to live young.

Birds are a group of animals that have feathers. They have two legs and two wings. Many lay eggs.

Most birds can fly. Birds also have beaks to help them eat food and carry things to their nests.

▲ **Ducks are birds. They lay eggs in nests.**

✓ **How are mammals and birds different?**

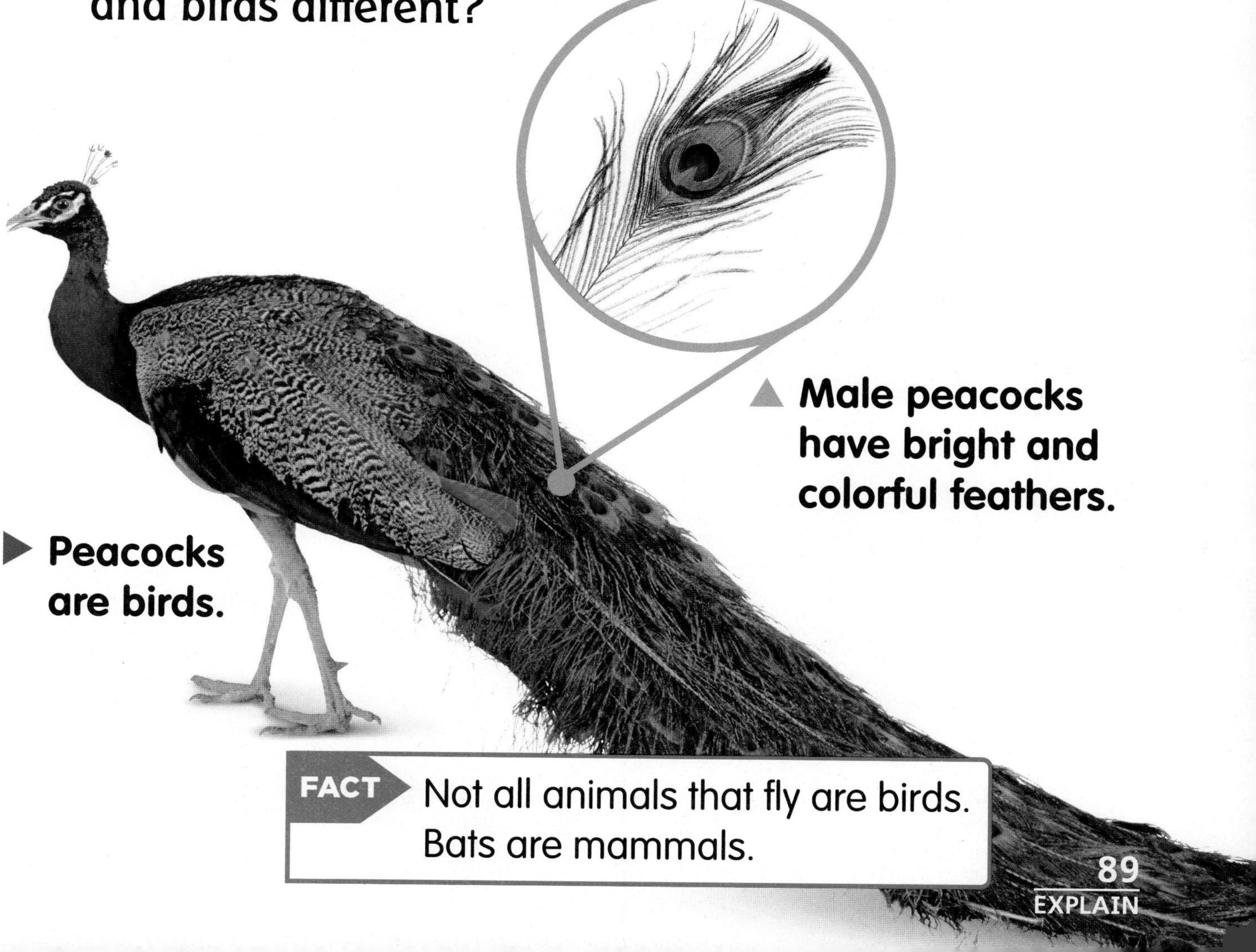

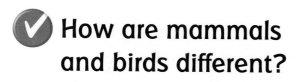

▲ **Male peacocks have bright and colorful feathers.**

▶ **Peacocks are birds.**

FACT Not all animals that fly are birds. Bats are mammals.

What are reptiles and amphibians?

Reptiles are a group of animals that have dry skin covered with scales. Some reptiles also have shells to keep them safe.

Most reptiles lay eggs. Snakes, turtles, and alligators are reptiles.

Amphibians are a group of animals that live on land and in water. Most amphibians have smooth, damp skin.

Quick Lab

Describe an animal. Have a partner guess the animal.

Amphibians usually hatch from eggs in water. Then they move to land when they are adults. Salamanders and frogs are amphibians.

✔ **Can you name some reptiles and amphibians?**

What are some other types of animals?

Fish are a group of animals that live under water. Most fish have scales.

Fish have fins that help them swim. Fish also have gills that help them breathe.

school of fish

Parts of a Fish

Read a Diagram

Why do fish have fins?

LOG ON *Science in Motion* Watch how fish move and breathe at **www.macmillanmh.com**

FACT Fish are not the only animals that travel in groups.

Insects are animals that have three body parts and six legs. Most insects lay eggs.

Ants and butterflies are both insects. Spiders are not insects. They have eight legs.

✔ **How do you know an ant is an insect?**

Think, Talk, and Write

1. **Compare and Contrast.** Compare different animals. How many legs do they have.
2. Write about why fish have fins.

Art Link

Use a hand lens to compare insects. Draw what you see.

LOG ON e-Review Summaries and quizzes online at www.macmillanmh.com

Lesson 2

What Animals Need to Live

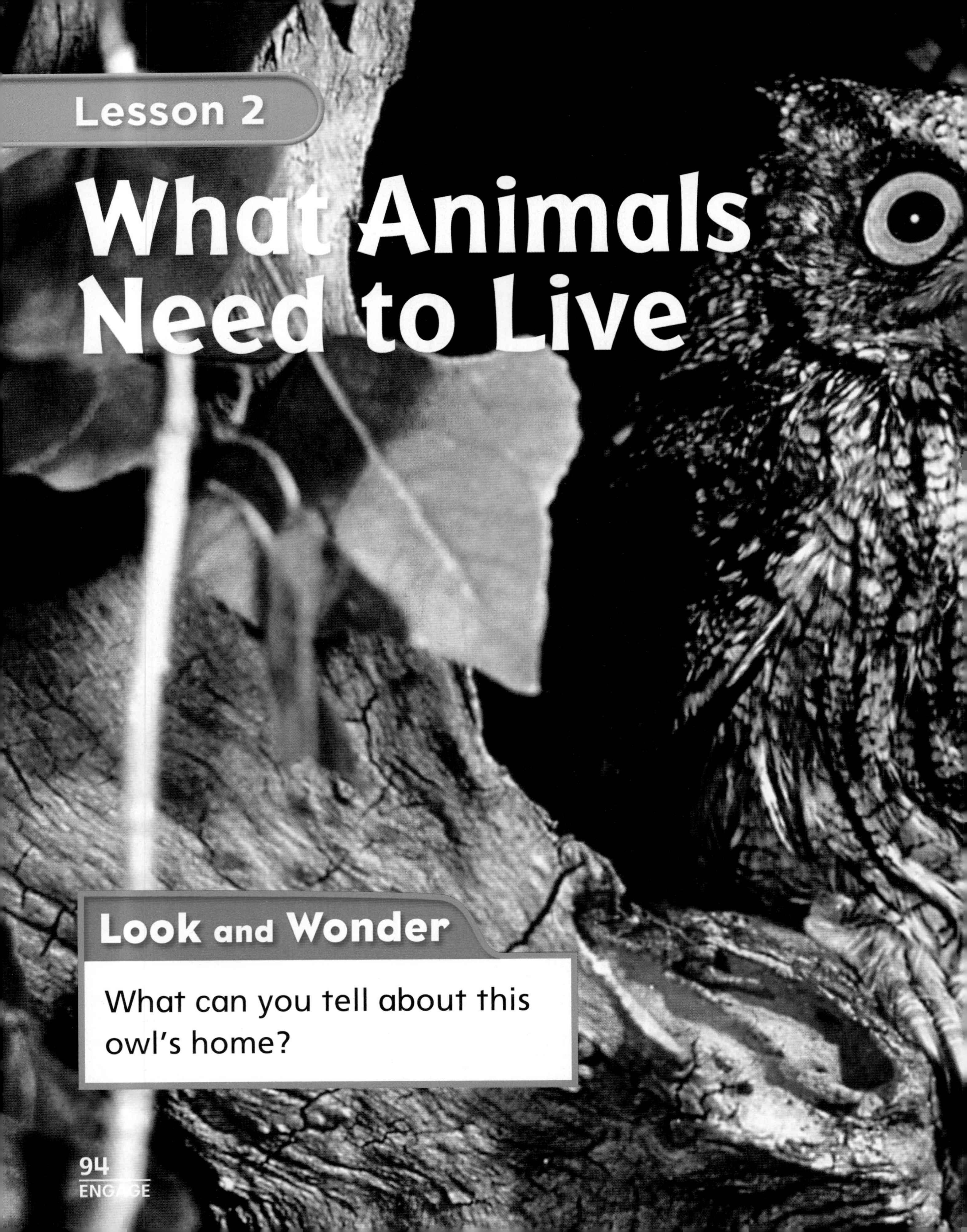

Look and Wonder

What can you tell about this owl's home?

Explore

Inquiry Activity

How do animals get what they need to live?

What to Do

1. **Make a Model.** Put fish food, water, and crickets in a terrarium.
2. **Observe.** Look at the crickets with a hand lens. How do they move? How do they get what they need to live?
3. **Communicate.** Draw a picture of your terrarium.

Explore More

4. **Compare.** Do all animals need the same things crickets need to live? How do you know?

You need

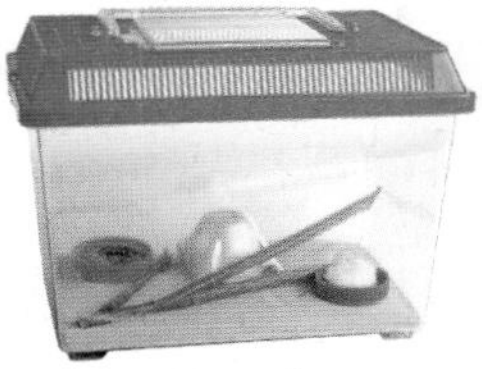
terrarium

water

fish food

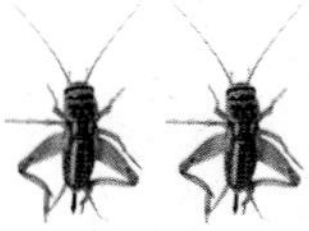
crickets

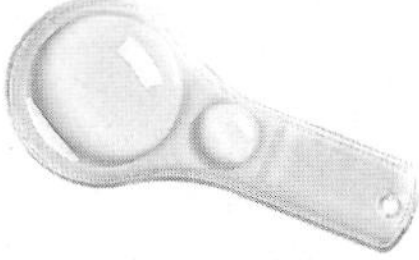
hand lens

Read Together and Learn

Vocabulary

shelter
gills
lungs

SCIENCE QUEST Explore what animals need with the Junior Rangers.

What do animals need to live?

Have you ever taken care of an animal? What did your animal need to live?

Animals are living things. Like you, they need food, water, air, and a safe place to live.

Zebras' Needs

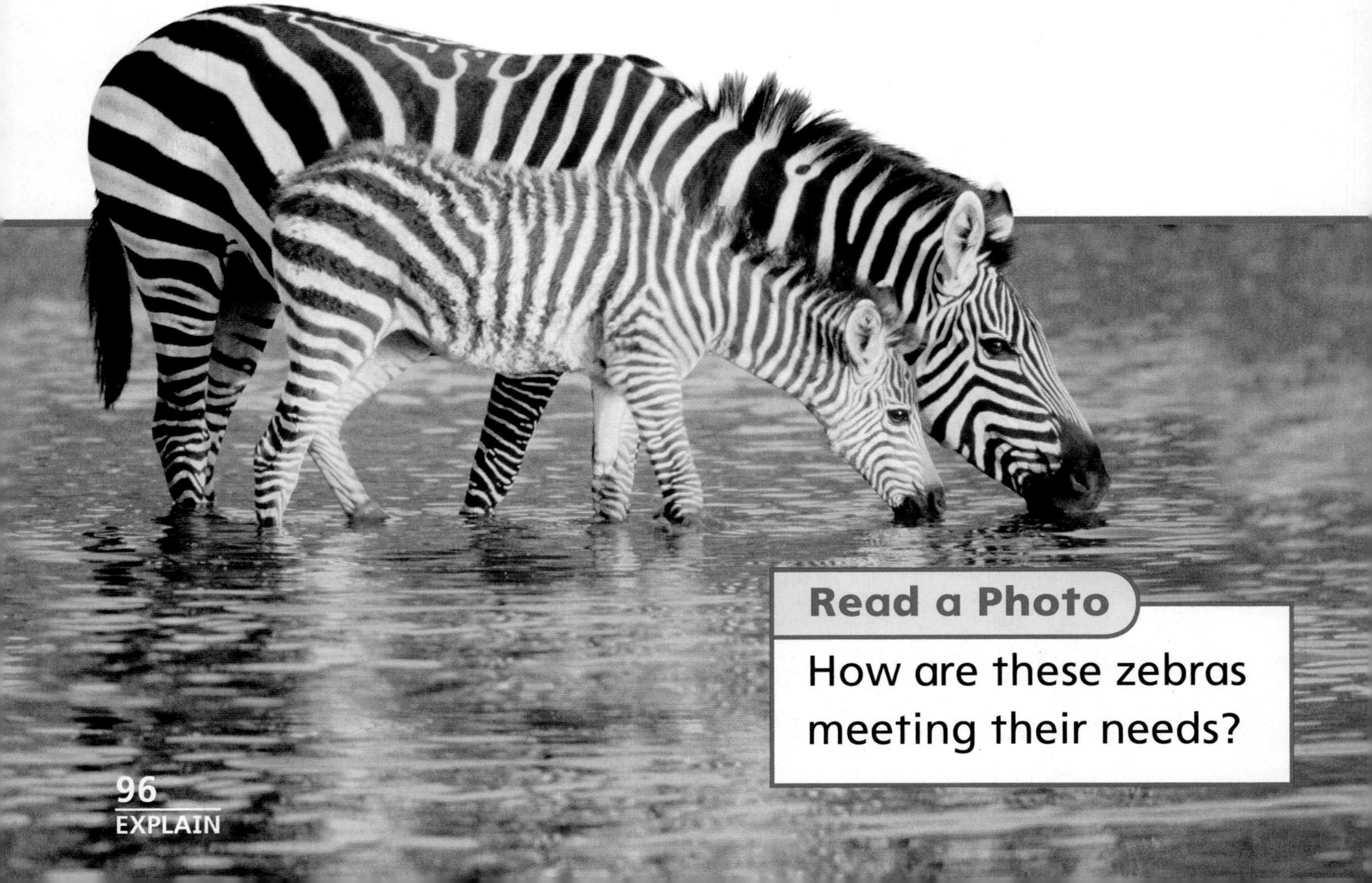

Read a Photo

How are these zebras meeting their needs?

Animals live in different kinds of places. Some animals live on land. Others live in water.

A **shelter** is a place where animals can live and be safe.

▲ **These raccoons find shelter in a log.**

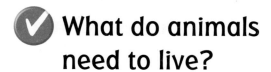

✔ **What do animals need to live?**

▲ **This bear eats a plant for food.**

How do animals meet their needs?

Animals have body parts to help them get what they need to live. Some use their eyes and noses to find food.

Eyes, ears, and noses also tell animals of danger. Legs, wings, and fins can help them get away from danger.

▲ **Wings help birds fly to find food. Their beaks help them eat food.**

Strong legs help mountain lions move fast to get food. Sharp claws help them catch their food. ▼

FACT Animals tell others where food is by moving their bodies, making sounds, or leaving smells.

Animals also have body parts that help them get air.

Gills help fish breathe in water. **Lungs** help other animals breathe air.

Quick Lab

Investigate how different body parts help animals meet their needs.

✔ **Which body parts help animals get food and air?**

▼ **Dolphins are mammals that live in the ocean. They have lungs.**

Think, Talk, and Write

1. **Classify.** How do body parts help birds and fish meet their needs?
2. Write about how you can help an animal meet its needs.

Health Link

How do you meet your needs?

LOG ON e-Review Summaries and quizzes online at www.macmillanmh.com

Animals' Needs

Esmeralda takes care of her cat. She makes sure it has what it needs to live.

Write a Story

Write about how Esmeralda cares for her cat. Tell how she helps her cat meet its needs.

Remember

Tell what happens first, next, and last.

LOG ON e-Journal Write about it online at www.macmillanmh.com

Animal Graph

Tom made a bar graph to show what kinds of pets his friends have at home.

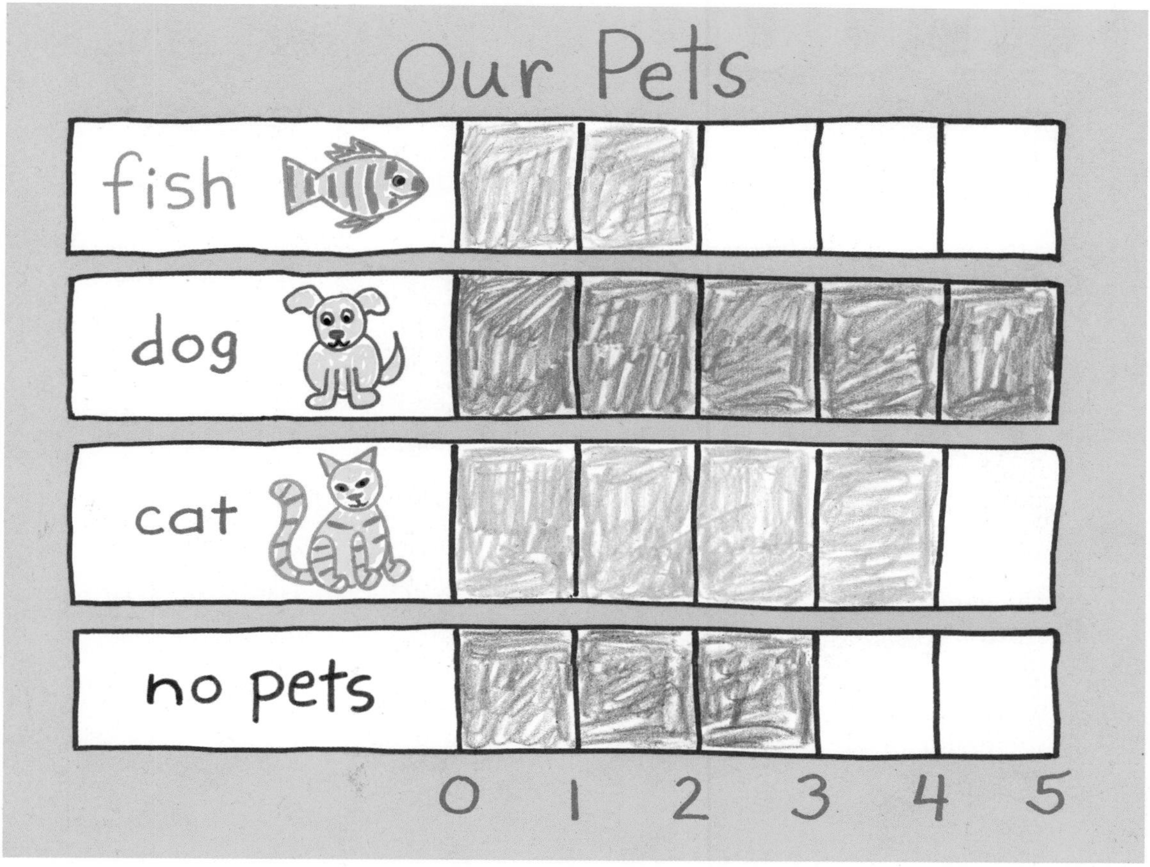

Make a Graph

Find out about your class's favorite pets. Make a bar graph to show which four pets your class likes best.

Remember

A bar graph helps organize information.

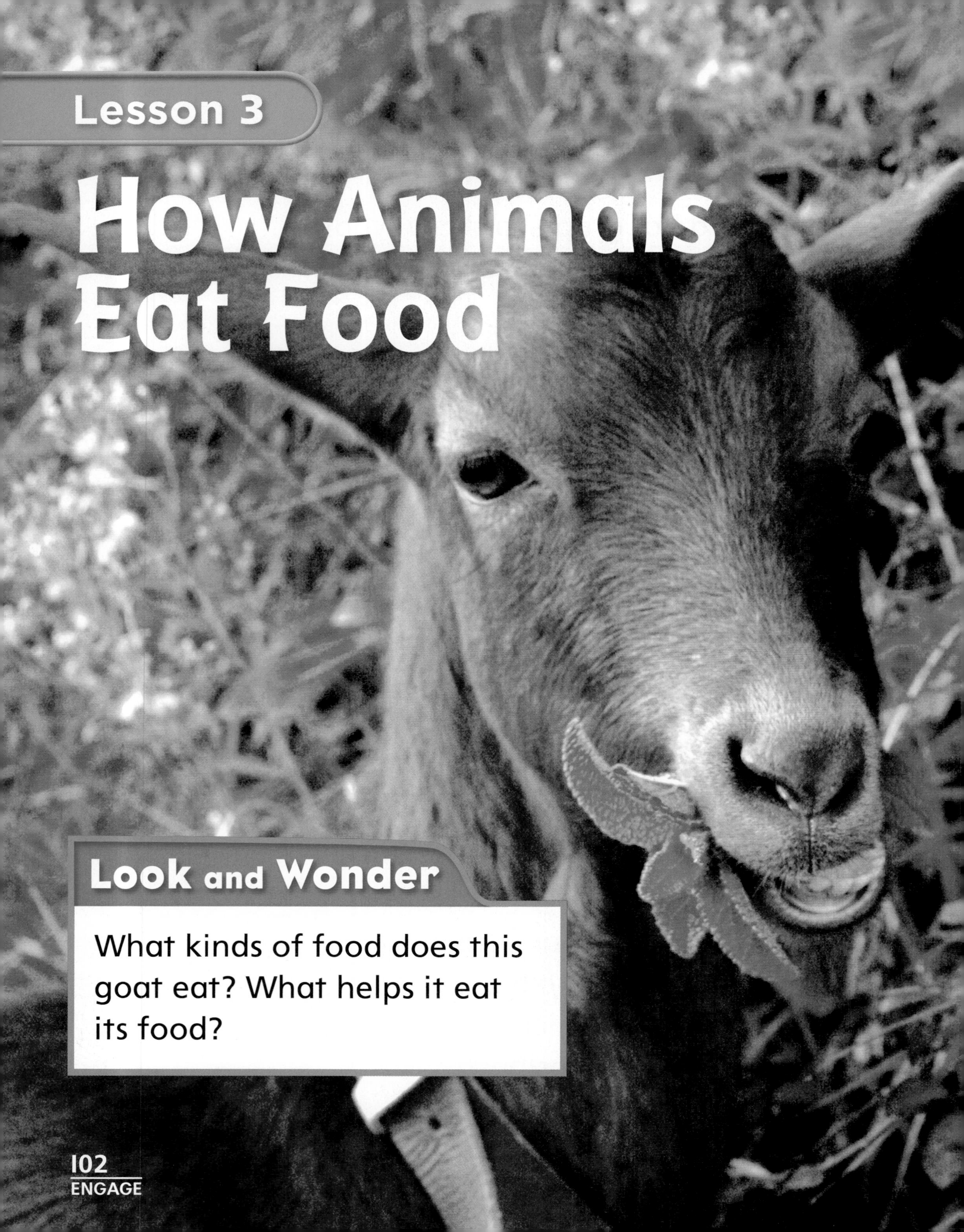

Lesson 3

How Animals Eat Food

Look and Wonder

What kinds of food does this goat eat? What helps it eat its food?

Explore

Inquiry Activity

How do teeth help you eat different foods?

What to Do

1. **Observe.** Try each type of food. Use a mirror to see which teeth you use.

△ **Be Careful.** Check with your teacher before eating any food!

2. **Record Data.** Draw and write about the teeth you used.

3. **Compare.** Look at the shape of your teeth. Why are they different?

Explore More

4. **Predict.** Which teeth would you use to bite a piece of meat? Why?

You need

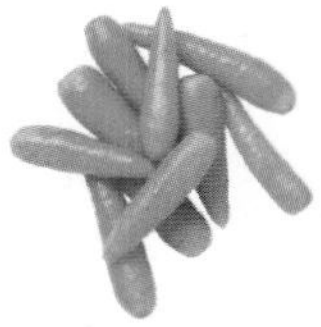

carrots

dried fruit

popcorn

mirror

Step 1

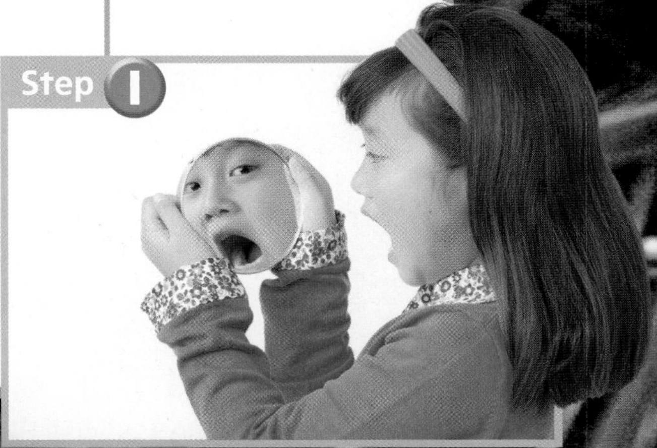

Read Together and Learn

Vocabulary

herbivore
carnivore

Which animals eat plants?

Animals eat food to get the energy they need to live.

Different animals eat different things. Some animals are herbivores.

horse

A **herbivore** is an animal that only eats plants. Horses and rabbits are both herbivores.

How do the flat teeth of a herbivore help it eat?

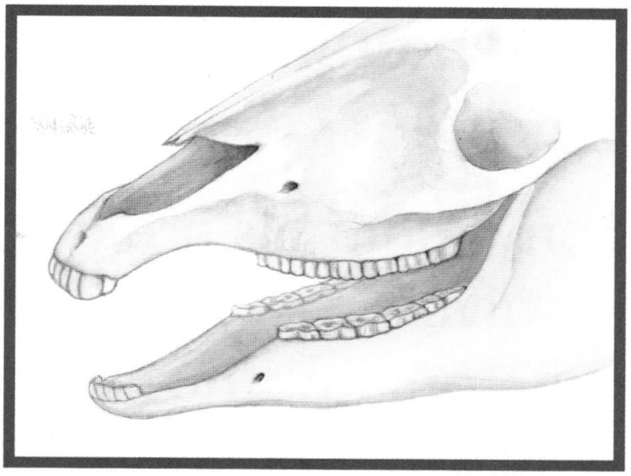

▲ **Herbivores have flat teeth to chew and grind plants.**

rabbit

Which animals eat meat?

Some animals are carnivores. A **carnivore** is an animal that eats only other animals.

Teeth

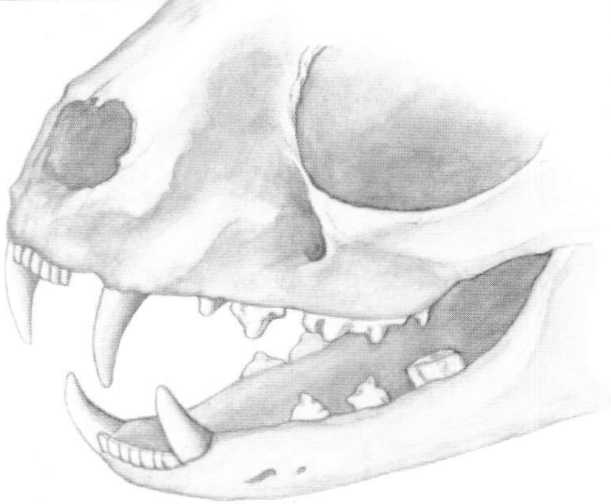

Read a Photo

How do these teeth help a carnivore eat?

Some carnivores have sharp claws to catch and hold their food. ▼

tiger

Carnivores have sharp teeth to rip and tear meat. Tigers and sharks are both carnivores.

Quick Lab

Compare what foods first graders like to eat.

✔ **How do sharp claws help some carnivores eat food?**

Think, Talk, and Write

1. **Put Things in Order.** Describe how a tiger first gets, then eats its food.
2. Write about and draw an animal that eats only plants.

Health Link

Why do you have to take care of your teeth?

LOG ON e-Review Summaries and quizzes online at www.macmillanmh.com

Lesson 4

Animals Grow and Change

Australia

Look and Wonder

What do you think the young kangaroo will look like when it gets older?

Explore

Inquiry Activity

How do animals grow and change?

1. **Observe.** Look at the pictures of different animals.
2. **Classify.** Sort the pictures below into two piles. Make one pile for adult animals and another pile for young animals.

3. **Compare.** How does each young animal change when it becomes an adult?

Explore More

4. **Infer.** What are different ways that animals can grow and change?

Read Together and Learn

Vocabulary

hatch

tadpole

How do mammals grow and change?

All animals grow and change. Animals are born, grow older, make other animals like themselves, and then die. This is their life cycle.

A life cycle is all of the parts of an animal's life.

This is the life cycle of a fox.

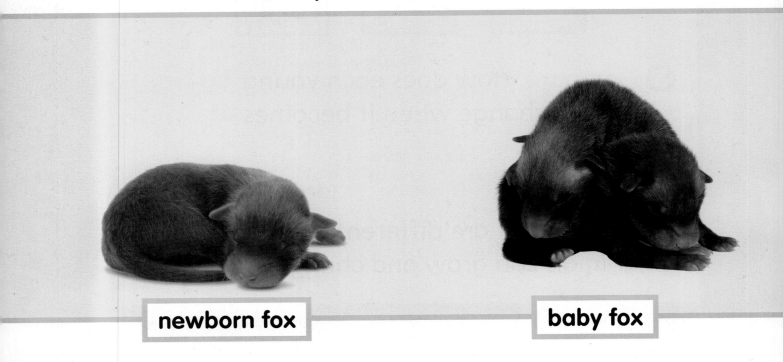

newborn fox

baby fox

Mammals grow and change their whole lives. They grow to look like their parents. Mammals give birth to live young.

When most mammals are born, they can not see or walk. Their mothers take care of them until they are old enough to take care of themselves.

✔ **How do mammals grow and change?**

young fox

adult fox

How do birds grow and change?

Birds are not born like mammals. They **hatch**, or break out of eggs.

When birds hatch, they can not fly. Parent birds take care of them. Parent birds find food for them.

Life Cycle of a Bird

▼ **This bird hatches from its egg.**

▼ **This bird is 3 weeks old. It has spiny feathers.**

Baby birds are not born with feathers. As they grow, their feathers grow in and cover their bodies.

Baby birds learn how to fly and find food by watching their parents.

Quick Lab

Draw and label your own life cycle. Start with yourself as a newborn.

✔ How do birds grow and change?

▼ **This bird is 5 weeks old. It is able to do more for itself.**

◀ **This bird is almost a fully grown adult.**

Read a Diagram

How has this bird changed since it hatched?

How do frogs grow and change?

A frog's life cycle begins in the water. Frogs lay eggs in water. **Tadpoles**, or young frogs, hatch from the eggs.

Tadpoles live in the water. They breathe with gills and have tails to swim.

▼ **This is the life cycle of a frog.**

As tadpoles grow older, they get ready to live on land. They grow legs to hop with and lungs to breathe air. Their tails become shorter.

Soon the tadpoles become adult frogs. They move onto land where they live most of their lives.

How do frogs grow and change?

adult frog

Think, Talk, and Write

1. **Compare and Contrast.** How is the life cycle of a frog like the life cycle of a bird? How is it different?
2. Write about how baby animals are alike and different from adult ones.

Art Link

Find out how a butterfly grows and changes. Draw its life cycle.

LOG ON e-Review Summaries and quizzes online at www.macmillanmh.com

Meet Melanie Stiassny

Melanie Stiassny is a scientist at the American Museum of Natural History. She studies fish. Some fish live in salt water. Some fish live in fresh water.

Melanie is an ichthyologist. That is a scientist who studies fish.

Some eels live in both salt water and fresh water. Baby eels hatch in the middle of the salty ocean. When they are big enough, they swim into a fresh water river to live.

Eels go back to the same part of the ocean where they hatched when they are ready to lay their eggs.

Baby eels hatch from the eggs. As they grow older, these eels begin the journey again.

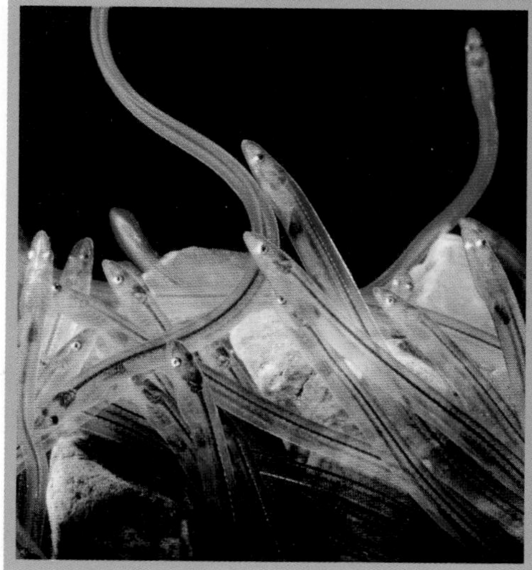

Baby eels hatch in the salty ocean.

Talk About It

Compare and Contrast How are baby eels like adult eels? How are they different?

American Museum of Natural History

I Read to Review

My Animal Book

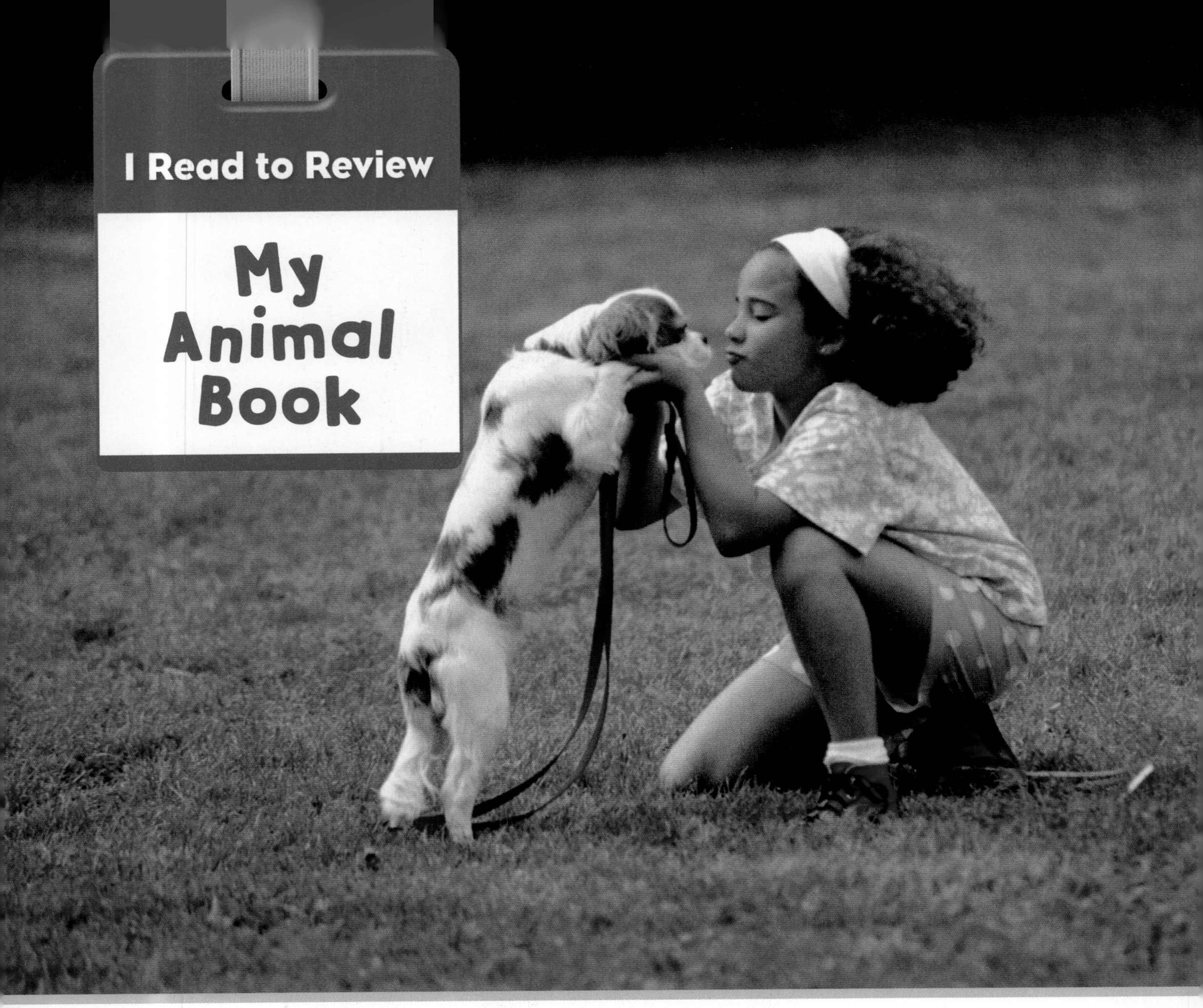

Do you have a pet
that needs your care?
Animals need food,
water, and air.

Do you know what
animals like to eat?
Some like plants.
Some like meat.

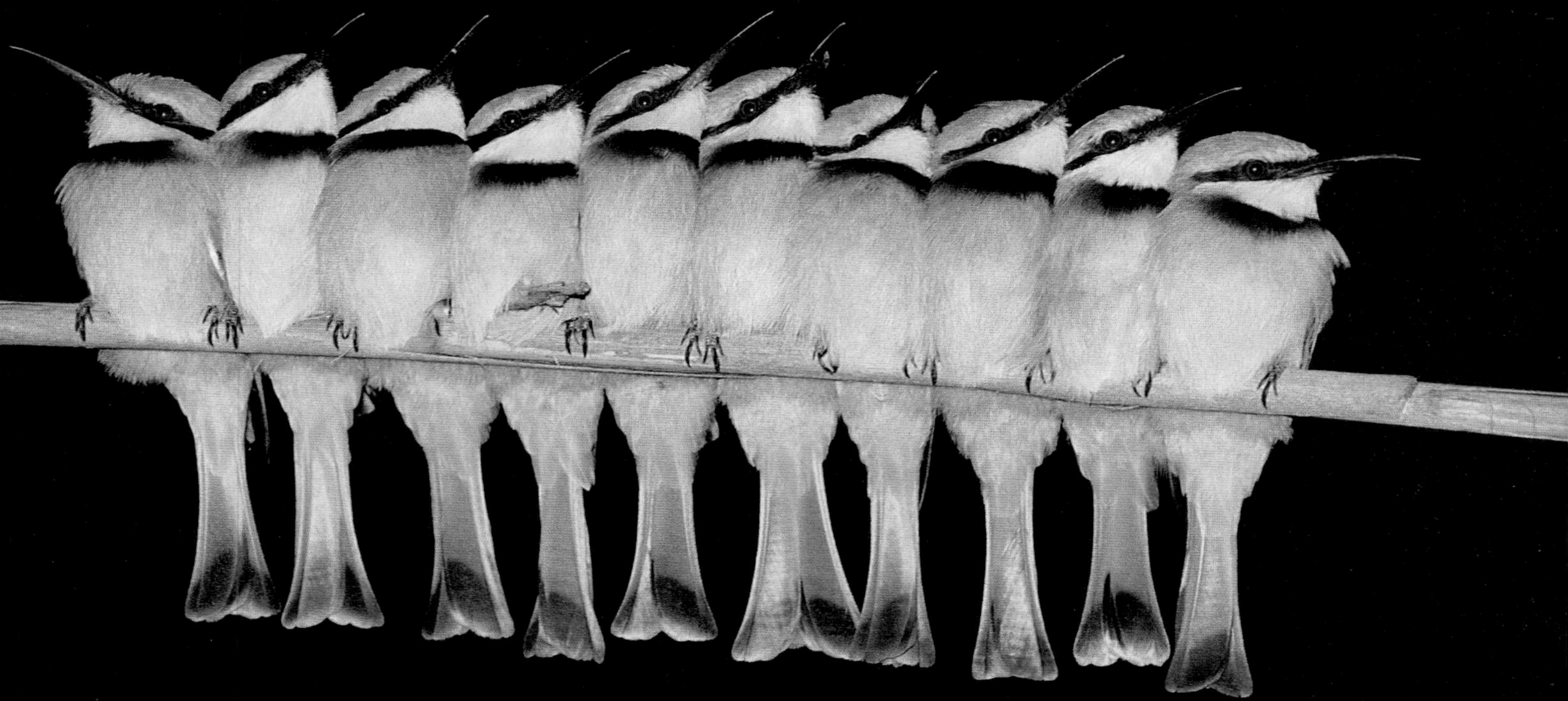

Do you like snakes?
They have scales.
Do you like birds?
They have feathers for tails.

Do you like horses?
They like to run.
Do you like lizards?
They lie in the Sun.

CHAPTER 3 Review

Vocabulary

What does each picture show?

amphibian
bird
fish
insect
mammals
reptile

1

2

3

4

5

6
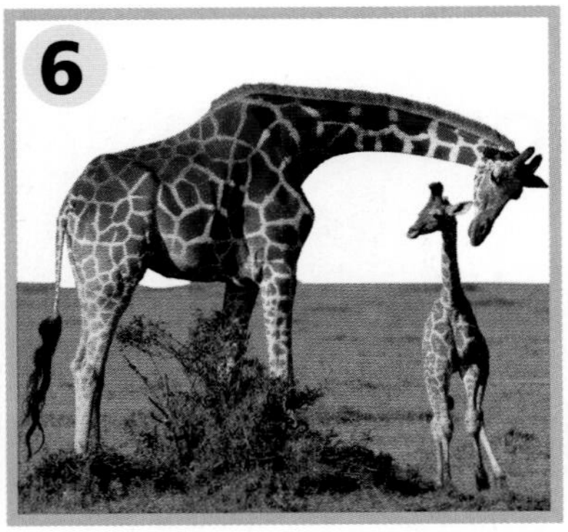

Answer the questions below.

7. How are baby mammals like their parents? How are they different?

8. **Compare.** What body parts do different animals use to meet their needs?

9. **Compare and Contrast.** Look at a bird and a butterfly. How are they alike? How are they different?

10. Tell how a frog egg becomes a frog.

The Big Idea

11. What do you know about animals?

CHAPTER 4

Places to Live

The Big Idea

Where do animals live?

China

Key Vocabulary

grassland a large open place with a lot of grass (page 128)

forest a place where there are many tall trees (page 130)

lake fresh water that has land all around it (page 134)

ocean salty water that is very large and deep (page 136)

More Vocabulary

habitat, page 128

adaptation, page 129

food chain, page 144

extinct, page 147

Lesson 1

Land Habitats

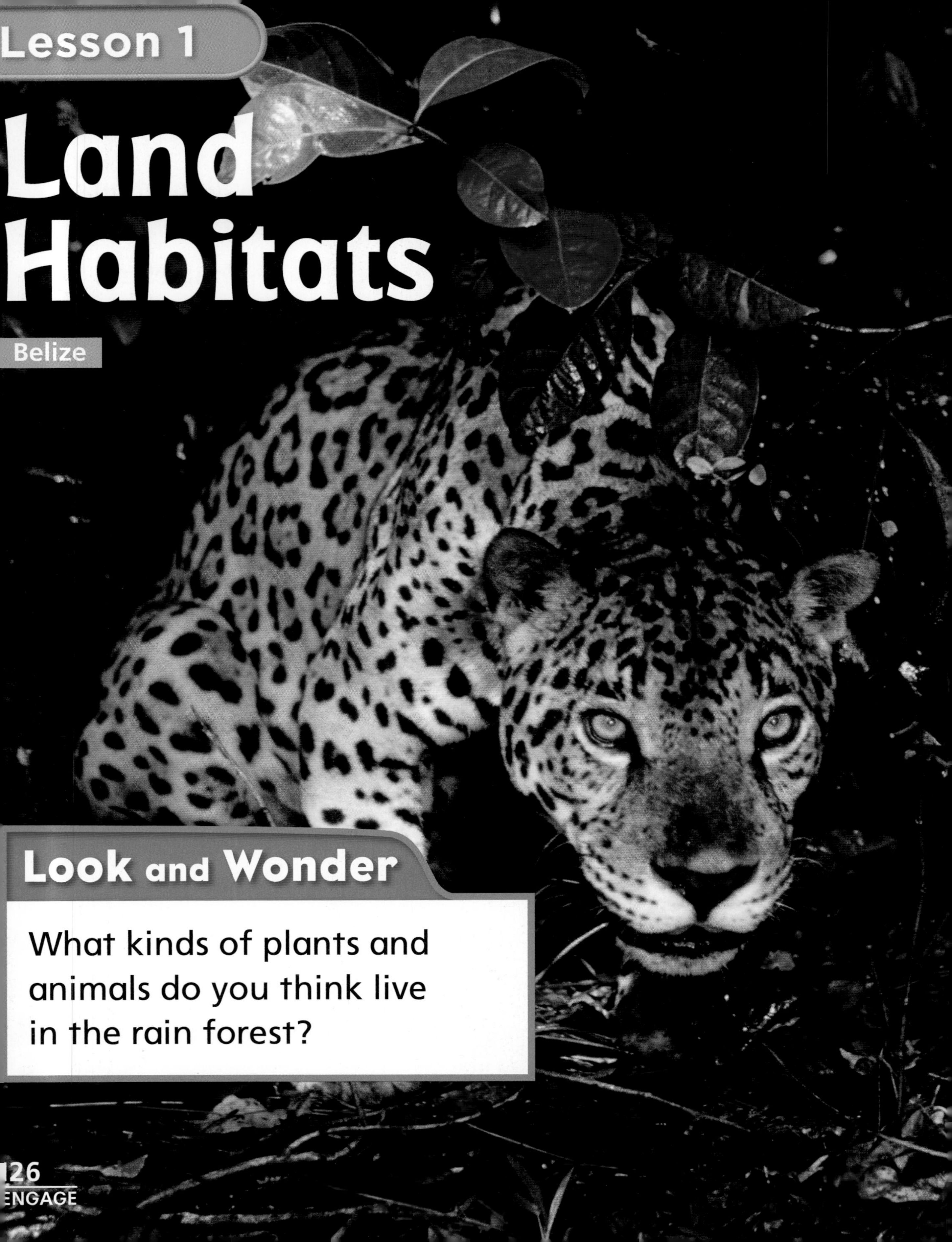

Belize

Look and Wonder

What kinds of plants and animals do you think live in the rain forest?

Explore

Inquiry Activity

Where do animals live?

What to Do

1. **Observe.** Look at pictures of animals and where they live.
2. **Communicate.** Tell a classmate what you notice about an animal and its home.
3. Draw and cut out pictures of an animal and its home.
   △ **Be Careful.** Remember scissors can be sharp.

Explore More

4. **Infer.** Why do you think different animals live in different places?

You need

magazines

crayons

scissors

glue

Step 3

Read Together and Learn

Vocabulary

habitat
grassland
adaptation
forest

SCIENCE QUEST Explore where plants and animals live with the Junior Rangers.

What is a grassland habitat?

Animals live in different places. The place where an animal lives is its **habitat**. One land habitat is a grassland.

A **grassland** is a dry place with a lot of grass. Some animals can hide in the grass. This helps them stay safe.

▲ **Prairie dogs live underground. They stay safe by hiding in holes from other animals.**

Animals have adaptations that help them stay safe in their habitats. An **adaptation** is a body part or a behavior that helps an animal survive.

Many grassland animals eat grass. They have flat teeth. Giraffes have long necks to eat leaves off tall trees.

✔ **How is a giraffe's long neck an adaptation?**

Giraffes' long necks help them spot animals that might eat them. ▼

What is a forest habitat?

A forest is another land habitat. A **forest** is a place where there are many tall trees.

Trees grow tall in forests to get lots of sunlight. Shorter plants can grow on the forest floor.

Quick Lab

Make a model of a forest.

▲ **Deer have good hearing, eyesight, and sense of smell to help stay safe.**

▲ **Raccoons have sharp claws to climb trees for food and safety.**

Some animals use the trees for food. Woodpeckers have long, sharp beaks to get food from inside a tree.

Other animals live in or store food inside trees.

Why is a forest a good place for animals to live?

Read a Photo

How is this plant helping the woodpecker meet its needs?

Think, Talk, and Write

1. **Predict.** What would happen if a forest were destroyed by a fire?
2. Write a list of reasons why it is good to be a deer or a raccoon in a forest.

Music Link

Listen to the song "Where Animals Live" on the Science Songs CD.

LOG ON e-Review Summaries and quizzes online at www.macmillanmh.com

Lesson 2

Water Habitats

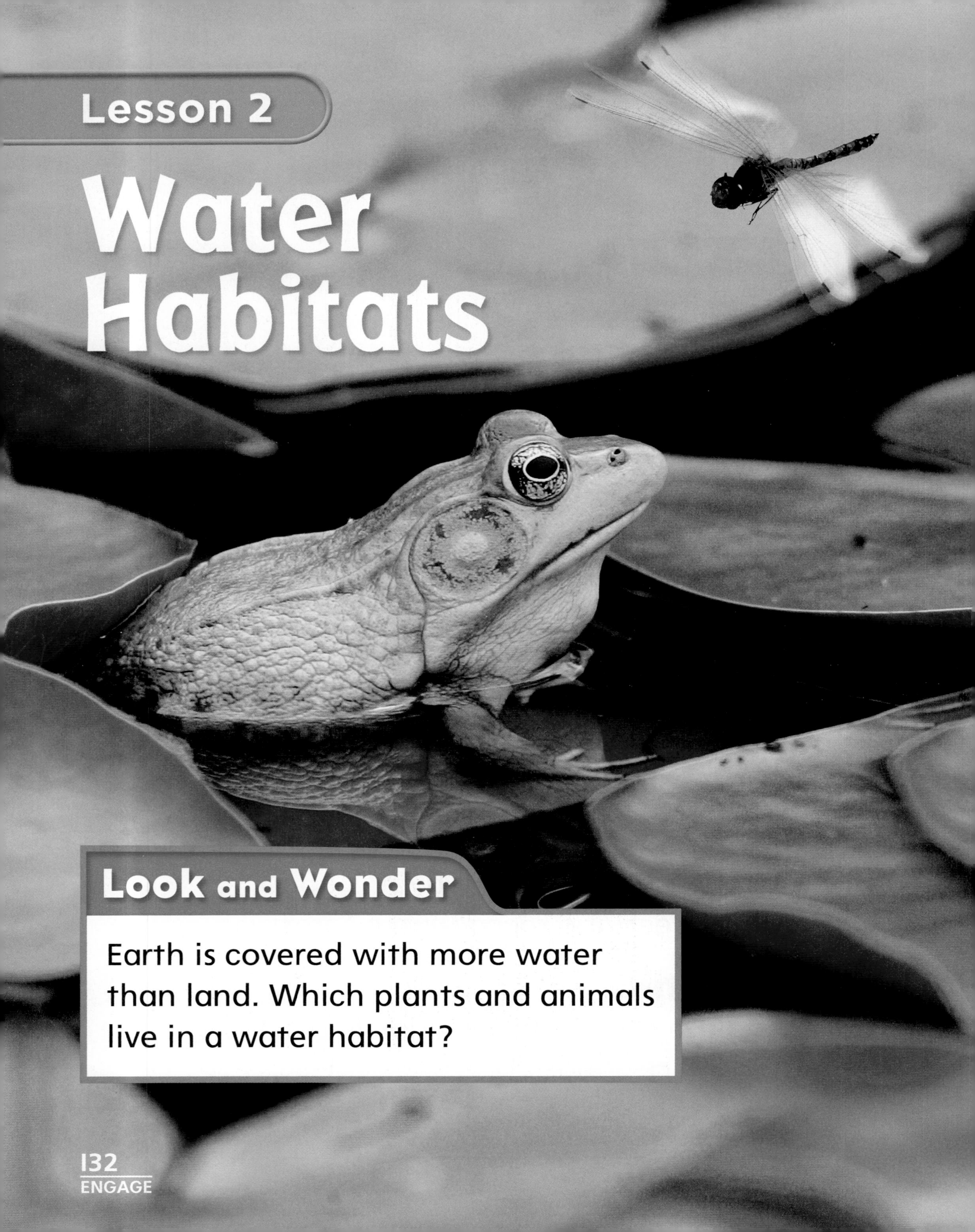

Look and Wonder

Earth is covered with more water than land. Which plants and animals live in a water habitat?

Explore

Inquiry Activity

How do plants and animals live in water?

What to Do

1. **Make a Model.** Put pebbles, a plant, water, and a fish in a clear tank.
2. **Observe.** Look closely at the fish with a hand lens. How do the parts of the fish help it live in water?
3. **Communicate.** Draw a picture of the aquarium.

Explore More

4. **Infer.** Could the fish live outside of water? Why or why not?

You need

clear tank

pebbles

plant

water

goldfish

Read Together and Learn

Vocabulary

lake

ocean

SCIENCE QUEST Explore plant and animal adaptations with the Junior Rangers.

What is a lake habitat?

A lake is one kind of water habitat.

A **lake** is water that has land all around it. Lakes are often fresh water. Fresh water has little or no salt in it.

Lake Habitat

Read a Photo

What living things live in this habitat?

Plants and animals live together in lakes. They need each other to survive.

Plants grow in and around the water. Animals use the plants for food and shelter.

Quick Lab

Build a dam in a tray with mud, sticks, and leaves.

How do plants help animals live in a lake?

▲ **Beavers use trees to build dams so they have a safe place to live.**

What is an ocean habitat?

An ocean is another kind of water habitat. An **ocean** is a very large and deep body of salty water.

Many mammals, fish, and plants live in oceans. They need each other to survive.

▲ **Fish have fins and tails to swim through water.**

Whales have thick skin to keep them warm in cold water. ▼

FACT Oceans are Earth's largest habitats.

Some animals, like whales, eat small fish in the ocean. Other ocean animals eat plants.

How are ocean animals alike? How are they different?

Think, Talk, and Write

1. **Main Idea and Details.** Describe a lake habitat.
2. Write about what helps a whale survive in an ocean.

Art Link

Find pictures of different whales. Compare them. Make an ocean collage.

LOG ON e-Review Summaries and quizzes online at www.macmillanmh.com

Arctic Adaptations

The arctic fox eats small animals. It can even find a lemming under the snow by listening carefully.

Its white fur helps it blend in with its surroundings. This helps the fox sneak up on animals it wants to eat.

Write About It

Write about how the arctic fox survives in the arctic.

Remember

Give information about one main idea.

LOG ON e-Journal Write about it online at www.macmillanmh.com

Count the Legs

Some animals use their legs to find food. Animals can have different numbers of legs.

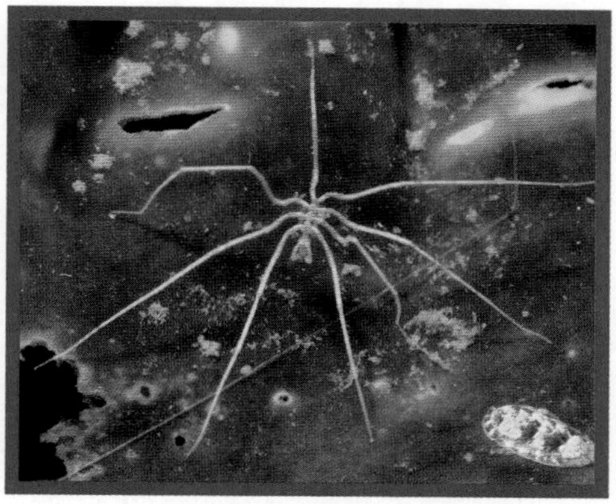

Put Them in Order

Count the number of legs each animal has. Put the animals in order from the smallest number of legs to the largest number of legs.

Remember

Start with the smallest number of legs.

Lesson 3

Plants and Animals Live Together

Look and Wonder

What kinds of animals live around you? What do they eat? Where do they get their food?

Explore

Inquiry Activity

Why are habitats important to animals?

What to do

1 Observe. What animals can you find near your school?

2 Communicate. Draw a picture of an animal and its habitat.

3 Infer. How does this habitat help the animal meet its needs?

Explore More

4 Draw Conclusions. Write about what you think would happen if the animal's food or habitat changed.

You need

paper

crayons

Step 1

Read Together and Learn

Vocabulary

food chain

extinct

Why do plants and animals live together?

Plants help animals live in their habitats.

Animals use plants for shelter and food.

This field mouse eats berries from a plant.

These skunks use a tree trunk as shelter.

Animals help plants live in their habitats, too.

Bees help plants make new plants. This happens when bees carry pollen from flower to flower. Pollen is powder inside a flower that helps makes seeds.

▲ **Pollen can stick to bees' legs.**

Why do plants and animals need each other in their habitats?

This spider uses plants to hold up its web. ▲

What is a food chain?

All living things need food. Food gives them energy to survive.

A **food chain** shows the order in which living things get food in their habitat.

Quick Lab

Draw a food chain where you are at the top of the chain.

Food Chain

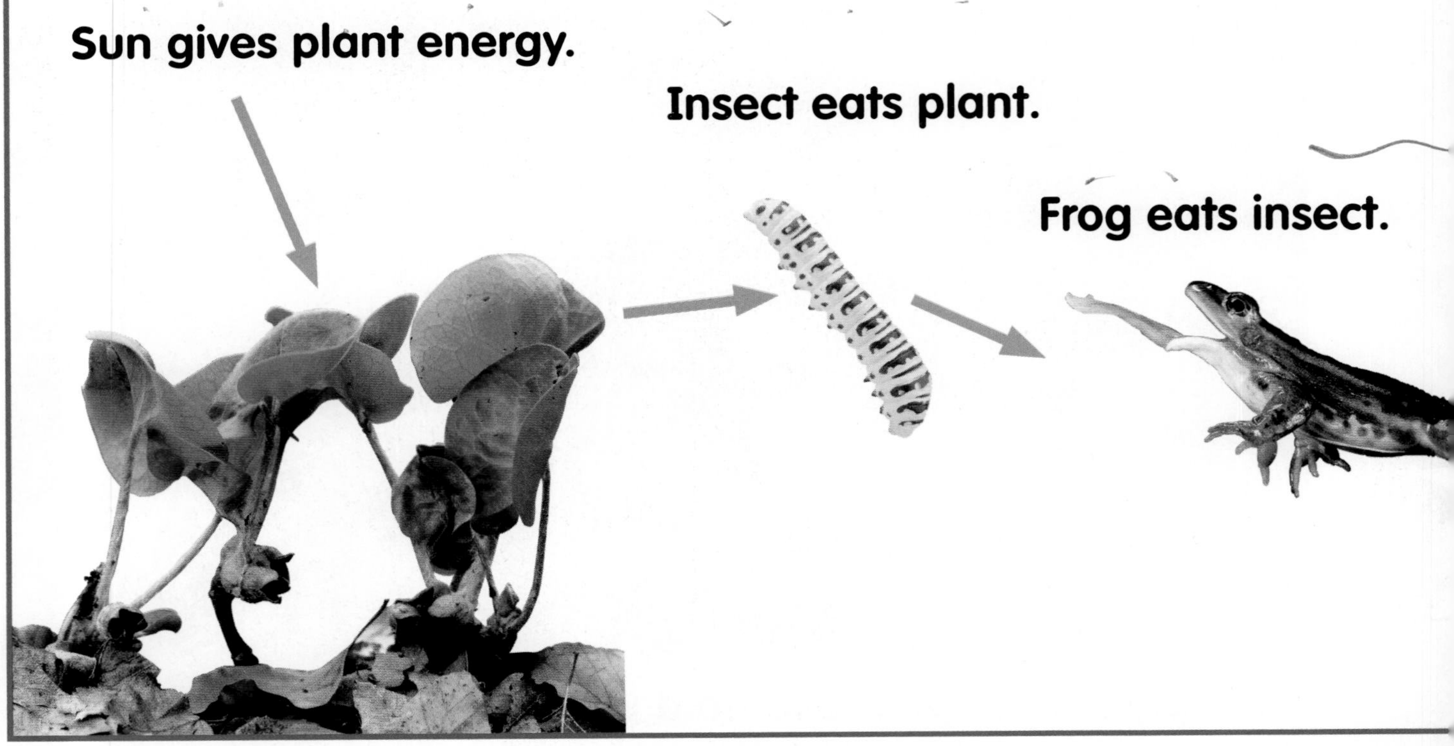
Sun gives plant energy.
Insect eats plant.
Frog eats insect.

The Sun is at the beginning of every food chain. Without the Sun, plants could not grow. Then there would be no food for animals.

Plants are the next link in most food chains. People are at the top, or end, of many food chains.

✔ **What kinds of food do you eat?**

Owl eats frog.

Read a Diagram

What is the first living thing in this food chain?

LOG ON *Science in Motion* Watch a food chain at **www.macmillanmh.com**

What happens to living things when a habitat changes?

People, animals, plants, and weather can all harm the living things in a habitat. They can cause the habitat to change.

Cutting down trees or polluting water can change habitats.

▲ **Lightning can strike a forest and start a fire.**

▲ **Plants and animals can be harmed and their habitats destroyed.**

When habitats change, animals and plants may not get what they need to live.

Some animals move to another habitat. Others might die. When all of one kind of plant or animal dies, it is **extinct**.

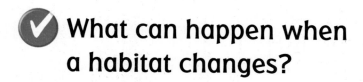

✔ What can happen when a habitat changes?

▲ **When their habitat changed, the woolly mammoth became extinct.**

Think, Talk, and Write

1. **Cause and Effect.** What would happen if all the trees in a forest were cut down?
2. Write about how plants help animals live in their habitats.

Art Link

Find out what different animals in the ocean eat. Draw an ocean food chain.

LOG ON e-Review Summaries and quizzes online at www.macmillanmh.com

Meet Jin Meng

Some animals, like *Tyrannosaurus rex*, lived long ago. How do we know what these animals ate?

Scientists like Jin Meng study fossils to find out. Fossils are what is left of living things from the past.

Jin works at the American Museum of Natural History. He looks closely at dinosaur fossils.

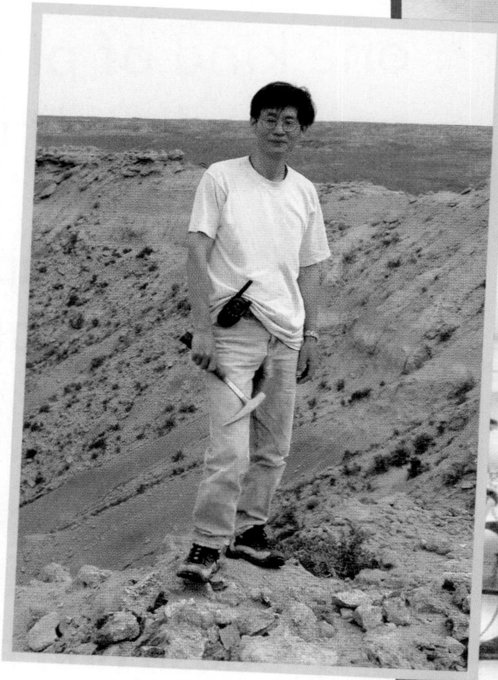

▲ **Jin Meng**

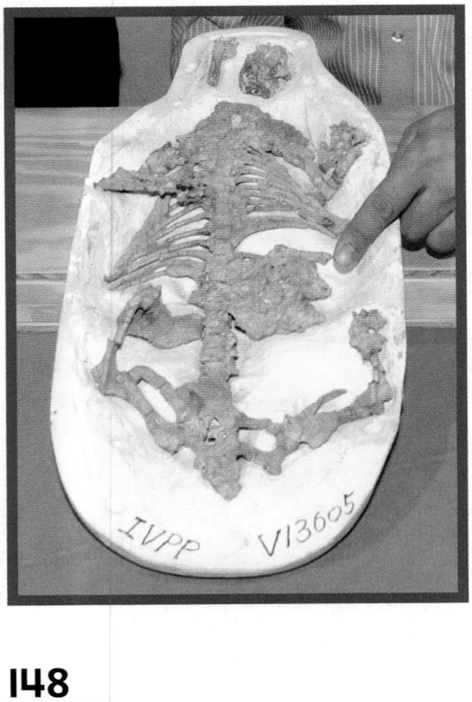

◀ **Small dinosaur bones found inside the stomach of this animal fossil tell us the animal ate meat.**

Scientists put bones together to help us learn what dinosaurs looked like.

Talk About It

Cause and Effect. Why can we not see a living *Tyrannosaurus rex?*

AMERICAN MUSEUM OF NATURAL HISTORY

I Read to Review

My Habitats Book

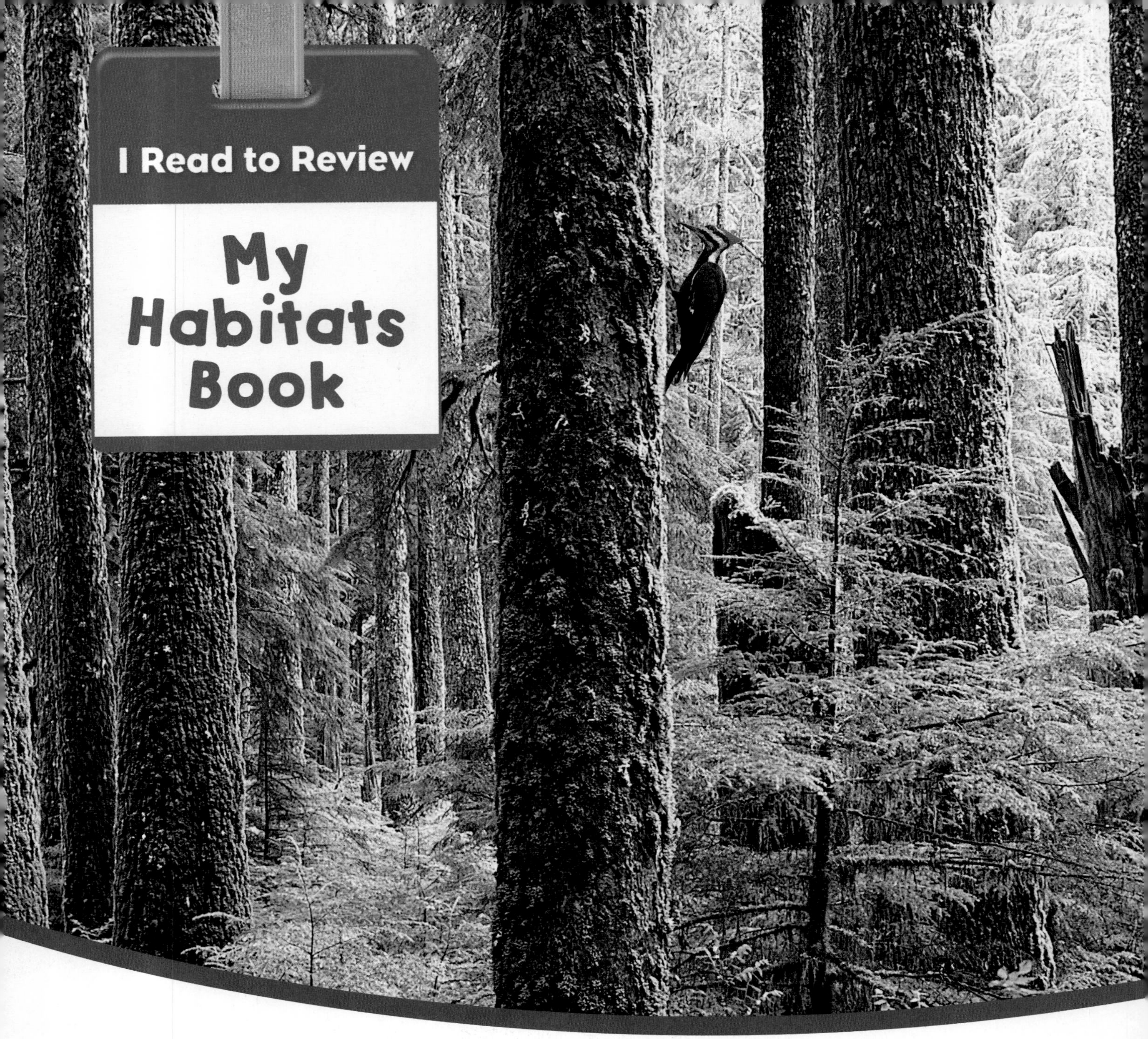

You can go to a forest
to see a woodpecker peck.

You can go to a grassland
to see a giraffe's long neck.

You can go to a lake
to see a snake with scales.

You can go to an ocean
to see a big whale.

CHAPTER 4 Review

Vocabulary

adaptation
food chain
habitat
ocean

Use each word once to complete the sentences.

1. The picture below shows a _____.

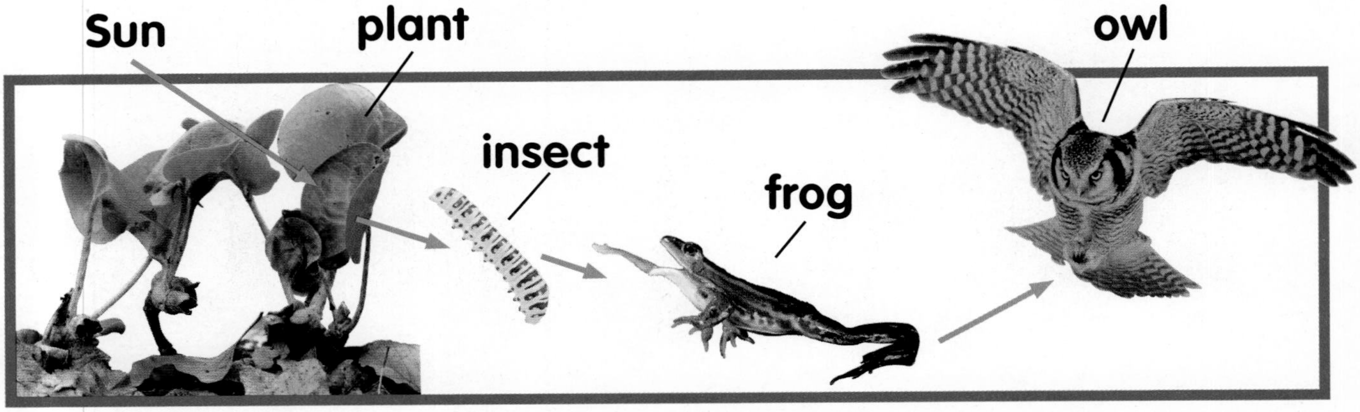

2. The largest habitat is an _____.

3. Giraffes have long necks to eat leaves from tall trees in grasslands. This is called an _____.

4. The place where an animal lives is called its _____.

Answer the questions below.

5. Describe a forest and a lake habitat. How have some animals adapted to survive in these habitats?

6. **Communicate.** Write about how plants and animals need each other to live in their habitats.

7. **Cause and Effect.** What might happen to an animal if its habitat changed?

The Big Idea

8. Where do animals live?

Veterinarian

Do you love taking care of animals? One day you could become a veterinarian. A veterinarian is a doctor for animals.

Many veterinarians care for pets. They help cats, dogs, birds, and other small animals. They also help cows, horses, and other big farm animals. Veterinarians need to study science so that they can take care of animals.

veterinarian

More Careers to Think About

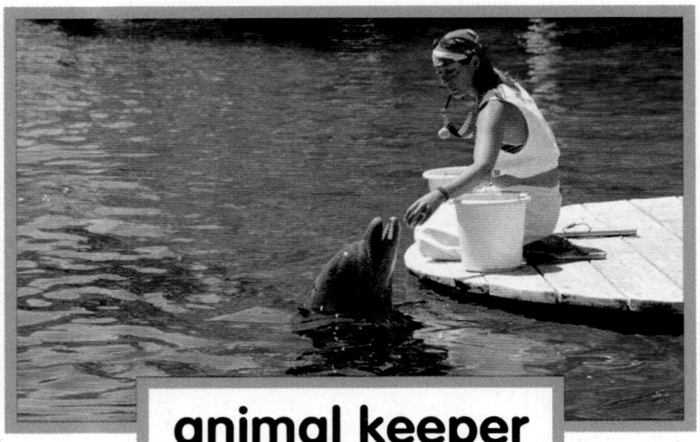

animal keeper

rescue worker

LOG ON e-Careers at www.macmillanmh.com

UNIT C

Our Earth

Water and ice can change the shape of rocks.

Bryce Canyon National Park in Utah

Literature
Poem

Sand

by Meish Goldish

Sand at the beach,
Sand at the shore.
Sand in the ocean
On the ocean floor.

Sand in the desert,
Sand on the ground.
Sand in a sandstorm
Blowing all around!

Sand from rock that has
Crumbled into grains.
Sand in a sand dune
Shaped by wind and rains.

Sand on an island,
Sand in the sea.
Sand in the sandbox
For you and me.

Talk About It

What makes sand? Where can you find it?

CHAPTER 5

Looking at Earth

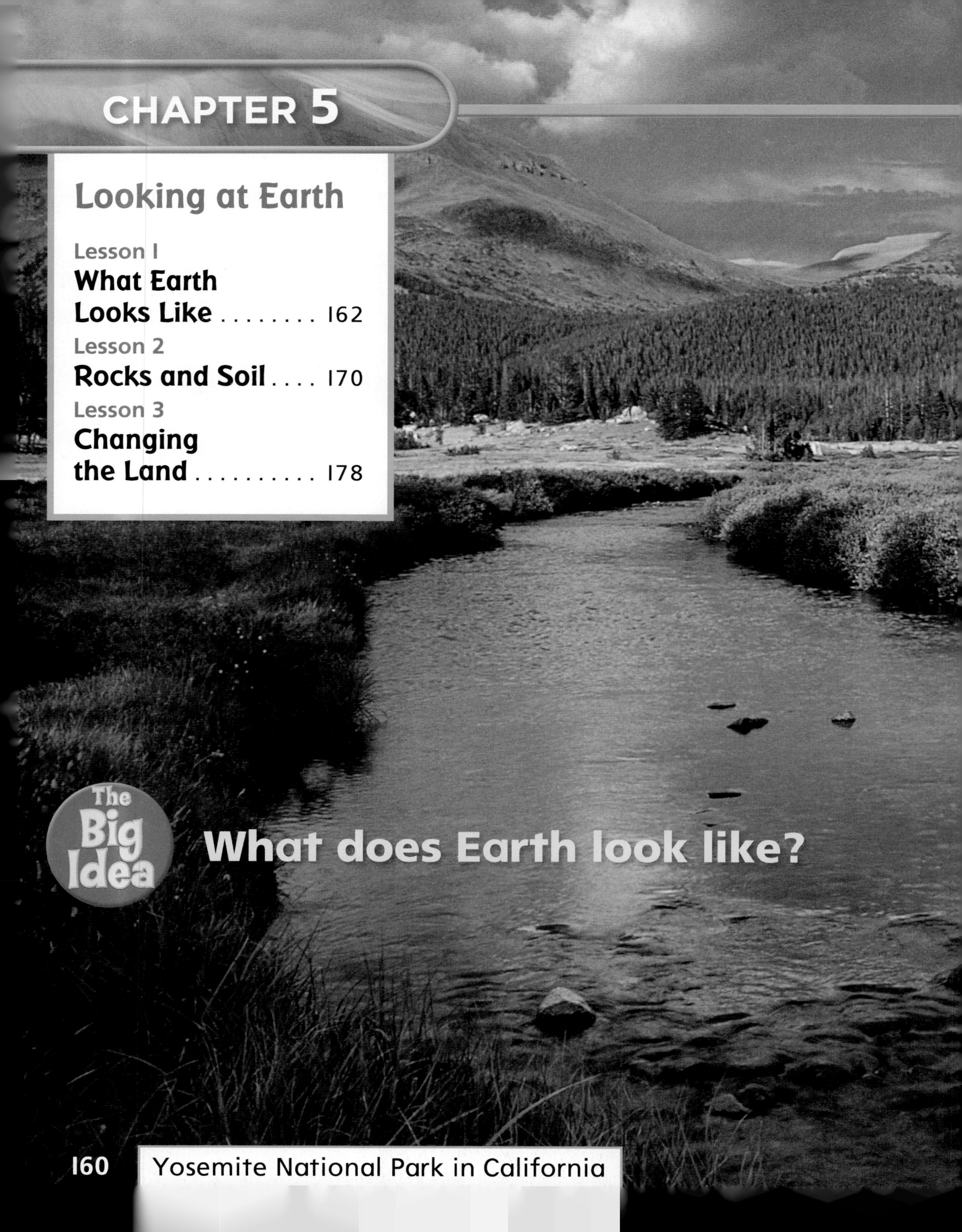

The Big Idea

What does Earth look like?

Yosemite National Park in California

Key Vocabulary

river fresh water that moves (page 167)

mountain land that is very high (page 168)

valley low land between mountains (page 168)

plain flat land that spreads out a long way (page 169)

More Vocabulary

continent, page 164

mineral, page 173

soil, page 174

weathering, page 180

erosion, page 182

Lesson 1

What Earth Looks Like

Moorea Island

Look and Wonder

An island is land surrounded by water. What does the land look like here?

Explore

Inquiry Activity

What can an island look like?

What to Do

1. **Make a Model.** Use clay to make an island in a clear bin. Add water.

   △ **Be Careful.** Remember to wash your hands!

2. **Observe.** Describe the land on your island. Is the land flat or high?

Explore More

3. **Communicate.** Write about the plants and animals that might live on or around your island.

You need

modeling clay

clear bin

Read Together and Learn

Vocabulary

continent
river
mountain
valley
plain

What is on the surface of Earth?

Earth is made up of land and water.

Land is the solid parts of Earth. There are seven large pieces of land on Earth. They are called **continents**.

Water surrounds the continents.
Water covers most of Earth.

Earth

Read a Photo

Where is the water on this photo of Earth?

✓ **What makes up Earth's surface?**

What is Earth's water like?

Not all water on Earth is the same. Most water on Earth is in salty oceans.

Many living things can not drink salt water. They need fresh water to survive.

Quick Lab

Draw the water in your area.

▼ **An ocean is a very large body of salt water.**

Fresh water is water without salt. Streams, rivers, and lakes can be made up of fresh water.

Streams flow downhill into rivers. **Rivers** may flow into lakes or oceans.

✔ **What are some different types of water?**

▲ **A lake is water that has land all around.**

▲ **A river can move fast.**

FACT Not all lakes are fresh water.

What is Earth's land like?

Not all land on Earth looks the same. Some land is high. Some land is low. Some land is flat.

A **mountain** is the highest type of land. Mountains come in all shapes and sizes. A **valley** is low land between mountains.

▲ **Plains are flat land that spread out a long way.**

✓ How are valleys and plains alike? How are they different?

Think, Talk, and Write

1. **Summarize.** What are some different kinds of land?
2. Write about how lakes and rivers are different.

Art Link

Look at a map. Find oceans, lakes, and rivers. Then draw the water you see.

LOG ON e-Review Summaries and quizzes online at www.macmillanmh.com

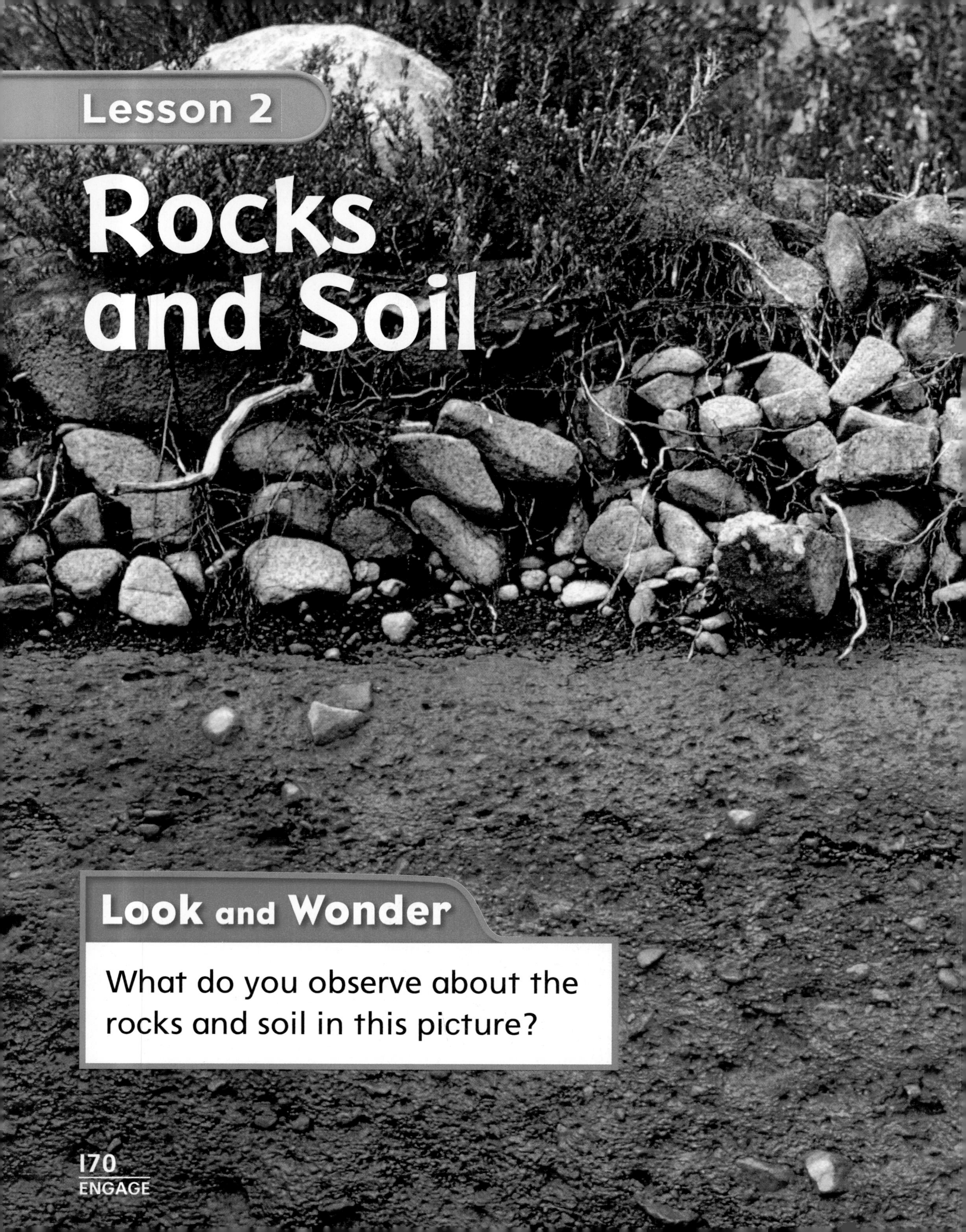

Lesson 2

Rocks and Soil

Look and Wonder

What do you observe about the rocks and soil in this picture?

Explore

Inquiry Activity

How can you classify rocks?

You need

rocks

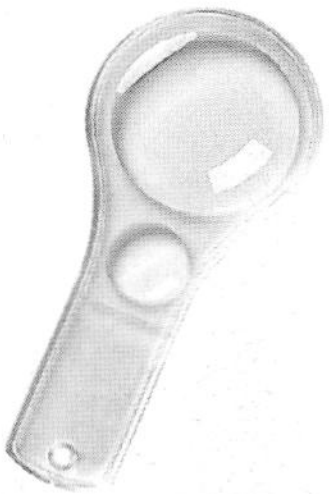

hand lens

What to Do

1. **Observe.** Use a hand lens to look closely at some rocks.
2. **Compare.** Do the rocks look or feel the same? How are they different?
3. **Classify.** Sort the rocks into groups. Explain your groups to a classmate.

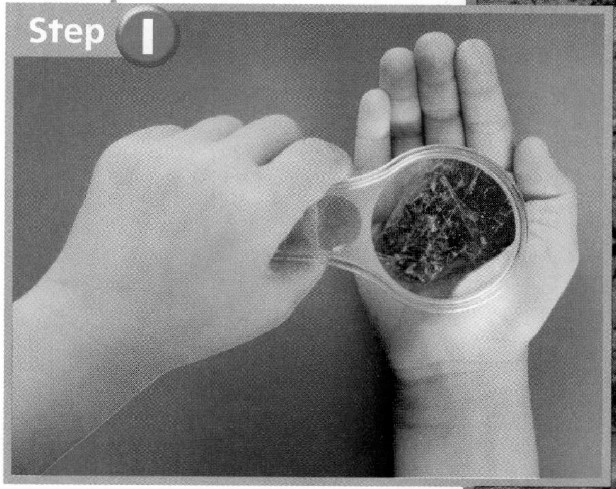

Explore More

4. **Make a Model.** Use clay to make a model of a rock. Include as many details as you can.

Read Together and Learn

Vocabulary

mineral

soil

What are rocks?

Rocks can have different shapes, sizes, and colors.

Some rocks are smooth. Other rocks are rough. Some are shiny. Others are dull.

Rocks

basalt

granite

sandstone

marble

All rocks are made of minerals. A **mineral** is a nonliving thing from the earth.

Some rocks are made of many minerals. Others are made of only one mineral.

✔ **How are all rocks the same?**

Quick Lab

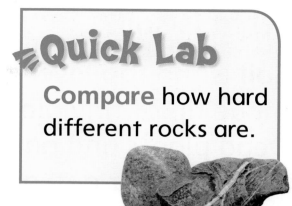

Compare how hard different rocks are.

chalk

pumice

limestone

obsidian

Read a Photo

How are these rocks alike and different?

FACT Not all rocks are hard.

What is soil?

Soil is the top layer of Earth. It is made up of tiny pieces of rocks. Dead plants and animals are also in soil. Water and air are, too.

◀ **Topsoil holds some water. It feels crumbly. Most plants grow well in topsoil.**

◀ **Clay soil holds a lot of water. It feels slippery. Many plants can not grow well in clay soil.**

◀ **Sandy soil holds a little water. It feels rough. Desert plants grow well in sandy soil.**

There are many different kinds of soil. Many soils are brown. Some are red, gray, or yellow.

Soils can have different plant, animal, or mineral parts in them.

▲ **Over time, this dead tree will become part of the soil.**

✓ What is soil made of?

Think, Talk, and Write

1. **Put Things in Order.** Take some rocks and put them in order from biggest to smallest.
2. Write about three different kinds of soil.

Social Studies Link

Collect some soil near school. Look at it with a hand lens and tell what you see.

LOG ON e-Review Summaries and quizzes online at www.macmillanmh.com

Meet Rondi Davies

Rondi Davies is a geologist at the American Museum of Natural History. A geologist is a scientist who studies rocks.

Rondi studies diamonds and how they were formed. Diamonds are made of carbon.

This is Rondi Davies.

uncut diamonds

Heat and pressure deep inside Earth can change carbon into diamonds.

If hot carbon cools very quickly, a diamond is formed.

If hot carbon cools slowly, graphite is formed. That is the gray tip of your pencil!

This is a cut diamond.

▲ graphite

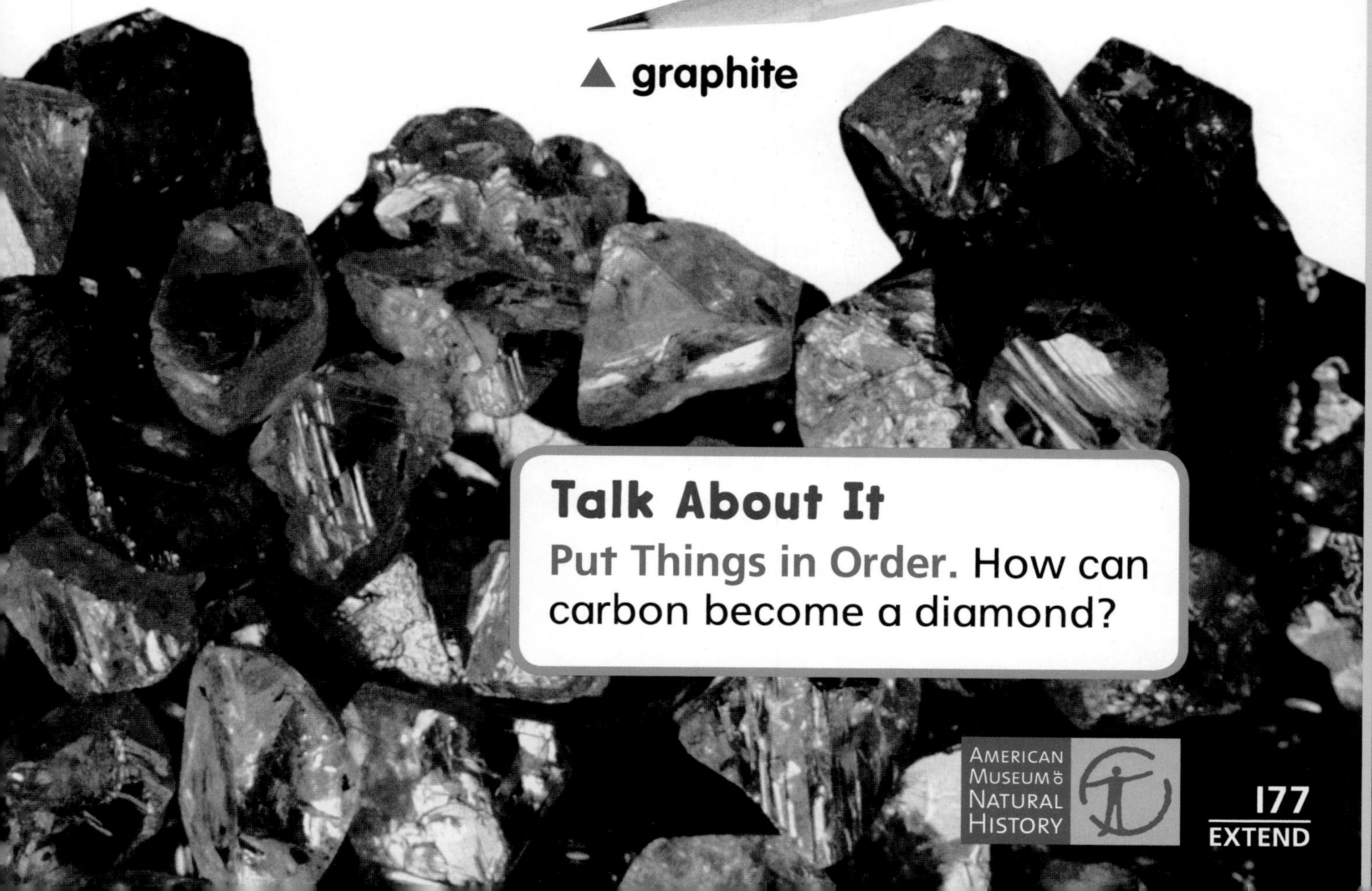

Talk About It

Put Things in Order. How can carbon become a diamond?

AMERICAN MUSEUM of NATURAL HISTORY

Lesson 3

Changing the Land

Columbia Glacier

Look and Wonder

This part of Earth is very cold. What is happening to the ground in this picture?

Plants can help prevent erosion. A plant's roots help hold soil in place so wind and water can not move it.

✔ **What is erosion?**

Quick Lab

Pour water over sand. **Observe** erosion.

sand dunes

Think, Talk, and Write

1. **Compare and Contrast.** Describe two ways water can change rock.
2. Write about how erosion could be slowed down.

Social Studies Link

After a rainy day, look for signs of soil erosion near you. How can you tell?

LOG ON **e-Review** Summaries and quizzes online at www.macmillanmh.com

Stopping Erosion

Look at the picture below. What do you think could be eroding the soil?

Write a Story

Write a story about what could help stop erosion in this picture.

Remember

A story has a clear beginning, middle, and end.

LOG ON e-Journal Write about it online at www.macmillanmh.com

Math in Science

Adding Rocks

Peter sorted his rocks into two groups. Then he wrote a number sentence to show how many rocks he had all together.

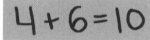

4 + 6 = 10

Write a Number Sentence

Collect your own rocks. Sort them into groups. Write a number sentence to show how many rocks you have.

Remember

A number sentence helps you solve a problem.

I Read to Review

My Earth Book

Earth is made of land and water. Where the ocean and land meet, there is sometimes sand.

Ocean waves hit land.

Small pieces of rock break off.

Sand is made of tiny pieces of rock.

Earth has different kinds of land.
Mountains are high land.
Valleys are low land.

Earth has different kinds of water.
Lakes have land all around.
Some plants grow in the water.

CHAPTER 5 Review

Vocabulary

Use each word once to complete the sentences.

mountain

river

soil

1. This moving water that flows into a lake or ocean is a _____.

2. Small rocks and dead plants and animals go into the _____.

3. This kind of land is called a _____.

Answer the questions below.

4. What are rocks made of? How can rocks change?

5. **Make a Model.** How could you make a model of a kind of land?

6. **Put Things in Order.** Put these rocks in order from smallest to biggest.

7. What are some ways water can change Earth?

The Big Idea

8. What does Earth look like?

CHAPTER 6

Caring for Earth

The Big Idea

Why do we need to care for Earth?

Key Vocabulary

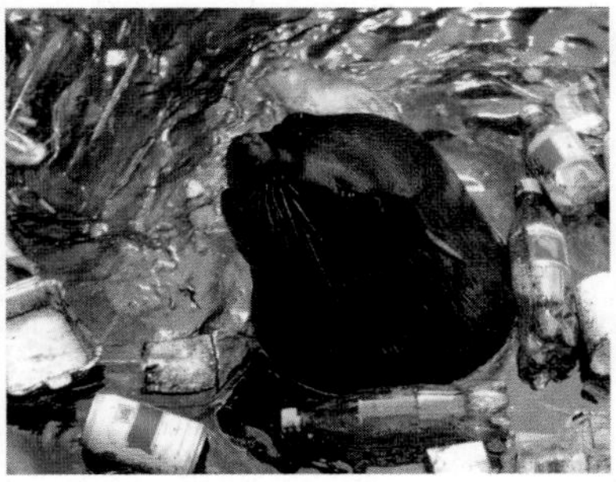

pollution harmful things in the air, land, or water (page 204)

reuse to use something again (page 210)

reduce to use less of something (page 212)

recycle to make a new thing from an old thing (page 213)

More Vocabulary

natural resource, page 196

conserve, page 210

Lesson 1

Earth's Resources

Look and Wonder

Many things we use come from plants, animals, or the Earth. Where do you think the things on this table came from?

Explore

Inquiry Activity

What things are made from plants or animals?

What to Do

1. Write the words "Plant" and "Animal" on sticky notes.
2. **Classify.** Look around the classroom. Put the sticky notes on objects that are made from either plants or animals.
3. **Communicate.** Make a list of objects that can come from plants or animals.

Explore More

4. **Investigate.** Pick an object from your classroom or house. Make a plan to find out what it is made from.

You need

sticky notes

Step 2

Read Together and Learn

Vocabulary

natural resource

What is a natural resource?

Things that come from Earth that people use are called **natural resources**.

We use living and nonliving resources every day. Animals, plants, rocks, soil, water, and air are some things we use.

◀ **You can use wood from trees to build things.**

People use natural resources in many different ways. We have to be careful not to use up all our natural resources.

✔ What natural resources do you use?

▲ You can use plants to make colorful dyes for many things.

▲ You can use wool from animals to make clothes.

You can use rocks to build walls, roads, and buildings.

Why is soil important?

Soil is a natural resource. It is very important for plants, people, and animals.

Plants grow in soil. People and animals can use these plants for food.

Quick Lab

Find out about an animal that uses soil for its home. Write about it.

Earth's Resources

Read a Photo

What resources can you see here?

LOG ON *Science in Motion* Watch how we use resources at **www.macmillanmh.com**

Soil can also be used to make things.

Clay is a kind of soil. Potters use clay to make objects such as cups.

How do living things use soil?

clay cup

Think, Talk, and Write

1. **Predict.** What plants do you think you will eat tomorrow?
2. Write a list of natural resources.

Music Link

Make up a song about things that grow in soil. Use the tune of "Old MacDonald Had a Farm."

LOG ON e-Review Summaries and quizzes online at www.macmillanmh.com

Lesson 2

Using Earth's Resources

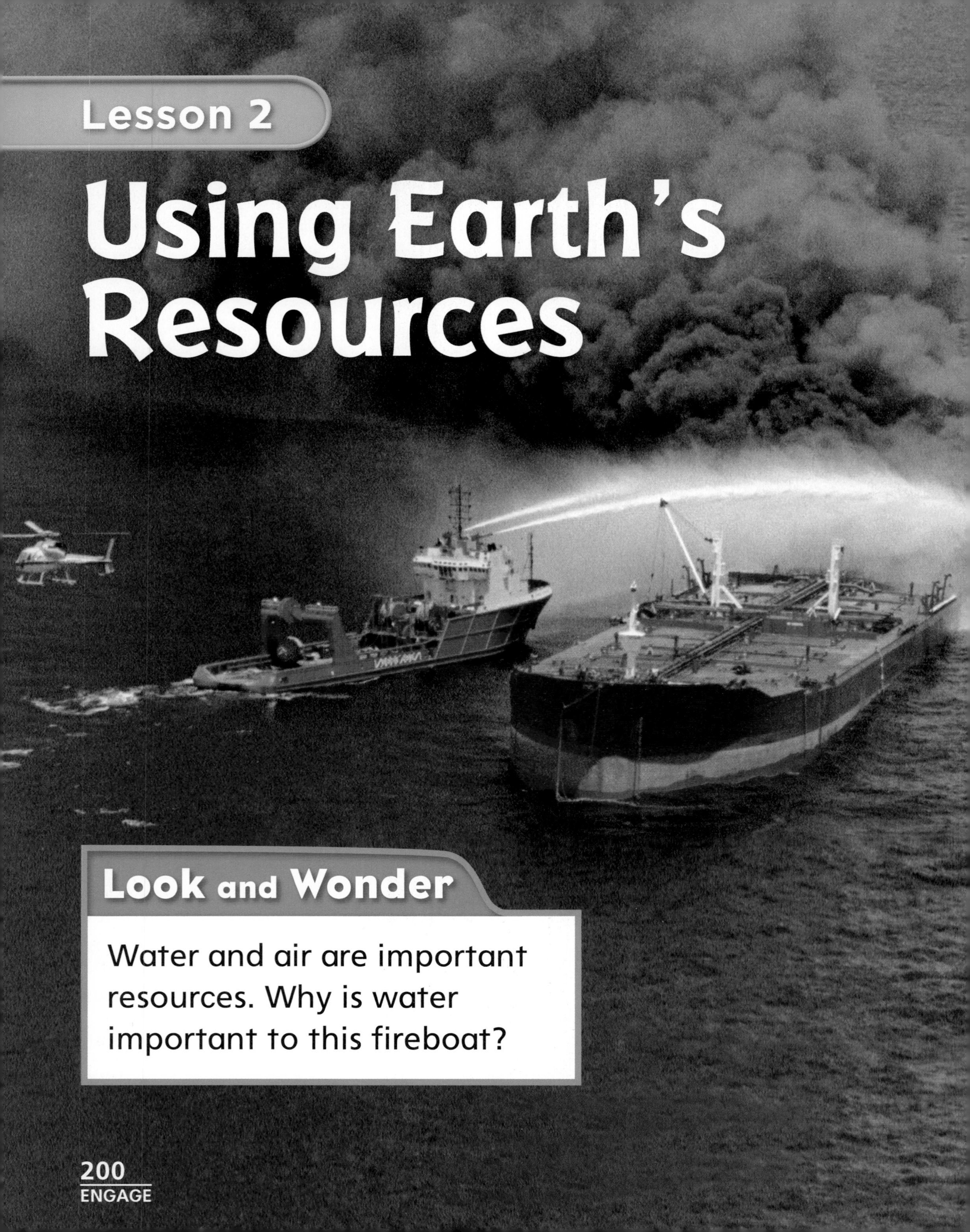

Look and Wonder

Water and air are important resources. Why is water important to this fireboat?

Explore

Inquiry Activity

When do you use water every day?

What to Do

1. **Investigate.** Find out ways you use water.
2. Make a class chart showing the different ways you all use water throughout the day.
3. **Record Data.** Use tally marks to record how many times water was used for each activity.

Explore More

4. **Infer.** Are there times when you could use less water?

You need

paper

crayons

Step 3

Read Together and Learn

Vocabulary

pollution

Why are water and air important?

All living things need water and air to live.

Plants need water to grow. We use water to wash and play. We also use it to cook and drink. People and animals need air to breathe.

Using Water

When water and air are not clean, we can not use them.

People need to keep water and air clean so all living things can use them.

✔ **Why is water important to living things?**

Read a Photo

How do people use water?

What is pollution?

Pollution is when there are harmful things in the land, water, or air.

Pollution happens when trash or dirt gets into the ground, water, or air.

Quick Lab

Use a sticky lid to catch what is in the air.

▼ **When people pollute water and land, it can hurt animals.**

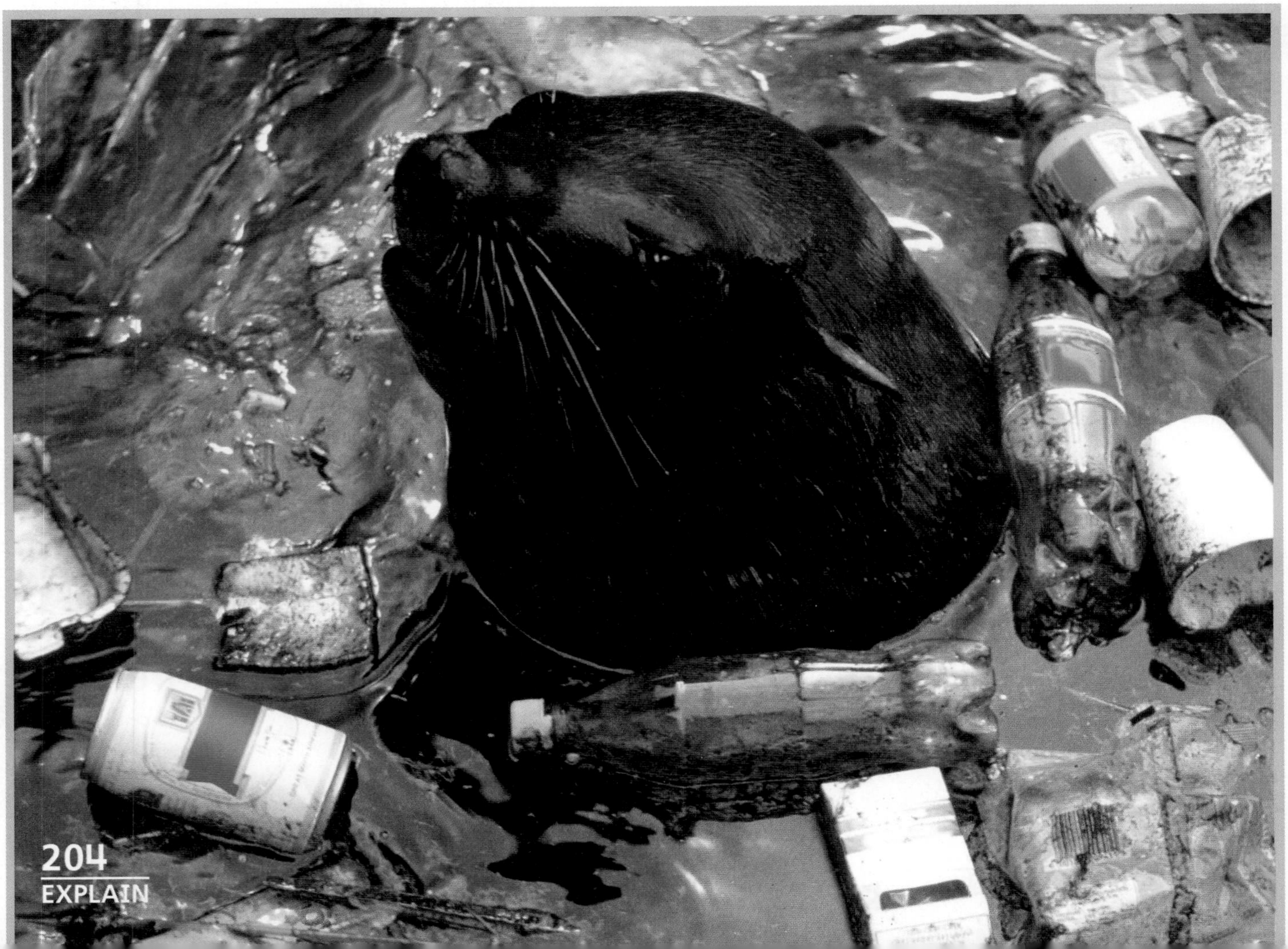

People and other living things can get sick if soil, water, or air are not clean.

There are many ways to keep Earth clean.

✓ **How can pollution be harmful?**

Think, Talk, and Write

1. **Cause and Effect.** What can happen if air gets polluted?
2. Write about how you use water.

Social Studies Link

Write about why you should not throw garbage on the ground or in the water.

LOG ON e-Review Summaries and quizzes online at www.macmillanmh.com

Meet Mark Siddall

It is important not to pollute Earth's waters. Many animals, like leeches, live in water. They need clean water to live.

Mark Siddall is a scientist at the American Museum of Natural History. He finds leeches in oceans, swamps, ponds, and streams.

Mark wants to know how many different leeches there are and how they take care of their young.

Mark is an invertebrate zoologist. That is a scientist who studies animals that do not have backbones.

▲ This is Mark in his lab.

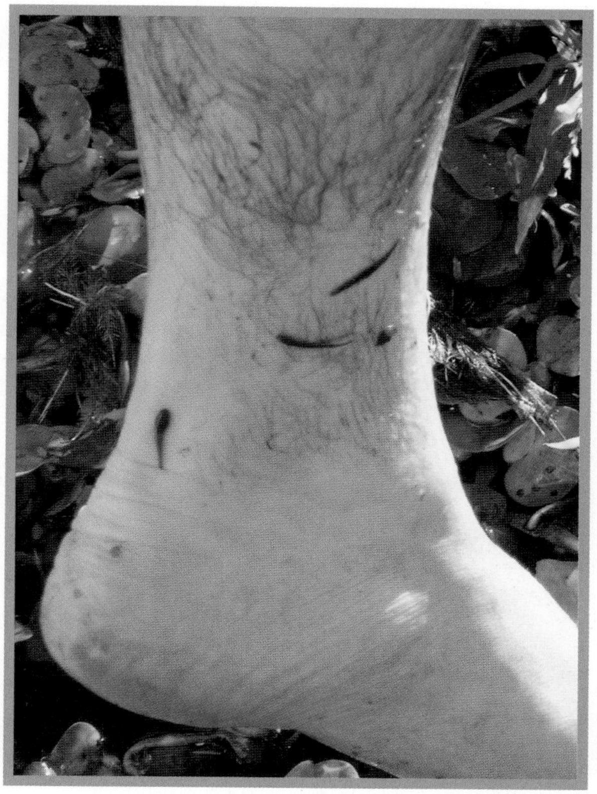

▲ Leeches suck blood to live. Mark uses his ankle as bait.

▼ This is a leech.

Talk About It

Problem and Solution. What could happen to leeches if people pollute Earth's water? What can we do to keep the leeches safe?

Lesson 3

Saving Earth's Resources

Look and Wonder

What do you think happens to the things you throw away?

Explore

Inquiry Activity

What happens to plastic when you throw it away?

What to Do

1. Put a piece of toilet paper into a cup of water. Put a piece of a plastic bag into another cup of water.
2. **Observe.** Leave the cups overnight. What happens to the paper and the plastic?
3. **Infer.** What would happen to a plastic bag and a paper bag if they were left outside on the ground?

Explore More

4. **Investigate.** Try this activity again using other materials.

You need

toilet paper

two cups of water

plastic bag

Read Together and Learn

Vocabulary

conserve
reuse
reduce
recycle

How can we reuse resources?

It is important to conserve natural resources. To **conserve** means to save, keep, or protect.

One way to conserve resources is to reuse things. To **reuse** means to use things again in a new way.

When you reuse things you do not have to buy new things. This keeps us from using up natural resources.

Quick Lab

Find a way to reuse something that you usually throw away.

✔ **Why is it important to reuse things?**

Reuse It

Read a Photo

What was reused to make this castle?

How can we save resources?

People can also save resources by reducing how much of a resource they use.

When you **reduce** what you use, you use less of it.

▲ **You can turn off a light when you leave a room to save electricity.**

▲ **You can shut the water off when you brush your teeth to save water.**

FACT Turning off the water each time you brush your teeth saves 38 juice boxes of water!

We can also recycle some materials so we do not have to use resources to make new ones.

To **recycle** means to make a new thing from an old thing. We can take used paper and recycle it into new paper!

Why is it important to recycle?

Many towns recycle paper, plastic, and glass. ▶

Think, Talk, and Write

1. **Problem and Solution.** How can we make sure we do not use up our natural resources?

2. Write about how you can reuse things in your classroom.

Social Studies Link

Find out what your community recycles.

LOG ON e-Review Summaries and quizzes online at www.macmillanmh.com

Saving Water

This girl is wasting water. She can save water by taking a drink and then turning off the fountain.

Write About It

Write about other ways people waste water. Tell what they can do to save water.

Remember

Use words to describe how to conserve water.

LOG ON e-Journal Write about it online at www.macmillanmh.com

Recycling Cans

John and his class picked up cans to recycle every day for one week. They made a picture graph to show how many cans they collected.

Cans We Collected
Days of week
Number of cans
Monday
Tuesday
Wednesday
Thursday
Friday

Read a Graph

On which day did John's class collect the most cans?

If the class got 5 cents for every can they collected, how much money would they have made on Monday?

Remember
You can use a graph to share your data with others.

I Read to Review

My Resources Book

We use natural resources every day.
We use water and air in lots of ways.

Soil is where plants can be found.
Some animals even live under ground.

We use rocks in many ways.
We even use them when we play.

We need to use resources with care.
We want them to always be there.

CHAPTER 6 Review

Vocabulary

Use each word once to complete the sentences.

conserve
natural resources
pollution
reuse

1. Water, rocks, and plants are all _____.

2. When our air and water get dirty, it is called _____.

3. When we use things over again we _____ them.

4. To save and protect our natural resources means to _____.

Answer the questions below.

5. How does soil help plants?

6. **Investigate.** How does your school reuse, reduce, or recyle? Make a plan to find out.

7. **Problem and Solution.** How could you reduce the amount of paper you use in school?

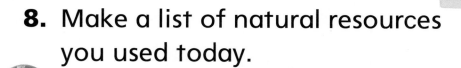

8. Make a list of natural resources you used today.

The Big Idea

9. Why do we need to care for Earth?

Hydrologist

Do you want to help care for Earth? A hydrologist is someone who studies water. Hydrologists study water in lakes, rivers, ponds, streams, and even under the ground.

Hydrologists test water to make sure it is clean and safe for plants and animals. They also study how water flows. They work to stop floods.

hydrologist

More Careers to Think About

gemologist

geologist

LOG ON e-Careers at www.macmillanmh.com

Weather and Sky

A rainbow never really touches the ground.

Literature

National Wildlife Federation

Ranger Rick

Magazine Article

This elephant knows just the thing for a hot day—a cool bath!

from *Ranger Rick*

Weather and ANIMALS

What do animals do in different kinds of weather? Let's find out!

Hot

The turtle lies in the sun to get warm.

▲ **turtle lying in the sun**

Cold

It is a cold, snowy day. The fox curls up. It uses its bushy tail as a blanket.

▲ **arctic fox curling up**

Rainy

Here comes the rain! The butterfly hides under a leaf. It hangs upside down to stay dry.

▲ **butterfly hiding**

Talk About It

Can all animals survive in all kinds of weather?

CHAPTER 7

Weather and Seasons

The Big Idea

What do you know about weather?

Key Vocabulary

spring the season after winter (page 242)

summer the season after spring (page 244)

fall the season after summer (page 250)

winter the season after fall (page 252)

More Vocabulary

weather, page 230

temperature, page 231

thermometer, page 232

rain gauge, page 232

wind vane, page 232

water vapor, page 236

cloud, page 237

season, page 242

Lesson 1

Weather All Around Us

Florida

Look and Wonder

What can you tell about the weather in this picture?

Explore

Inquiry Activity

What can you observe about the air?

What to Do

1. Use a craft stick and a piece of streamer to make a weather tool.
2. **Predict.** What kind of weather do you think this tool will be able to tell you about?
3. **Observe.** Take your weather tool outside and hold it in the air. What happens?

Explore More

4. **Compare.** Use your weather tool to test the air on different days. What do you notice?

You need

craft stick

streamer

tape

Read Together and Learn

Vocabulary

weather
temperature
thermometer
rain gauge
wind vane

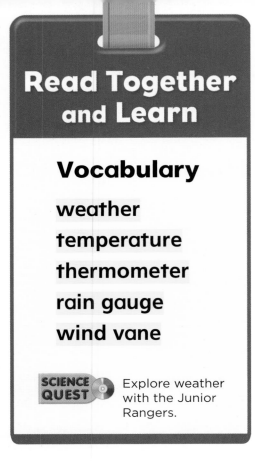

SCIENCE QUEST Explore weather with the Junior Rangers.

What is weather?

You probably think about the weather every day. Weather affects what you wear and what activities you can do.

Weather is what the sky and air are like each day. The sky might be sunny or cloudy. The air might be rainy, snowy, or dry.

sunny

cloudy

Weather changes from day to day. The Sun warms the air, changing the temperature. **Temperature** is how warm or cold the air is.

Moving air is called wind. Wind can move slowly or very fast.

How does temperature affect you?

rainy

snowy

How can you measure weather?

You can use tools to measure weather. Different tools measure different weather.

Quick Lab

Use a thermometer to **compare** the temperature inside and outside.

Weather Tools

◀ **A thermometer measures temperature.**

A rain gauge measures how much rain falls. ▶

◀ **A wind vane shows the direction of the wind.**

Read a Photo

Which of these tools measures rainfall?

Some tools measure temperature. Other tools measure wind or rain.

✓ Why is it useful to measure weather?

This child is measuring rainfall with a rain gauge. ▶

Think, Talk, and Write

1. **Put Things in Order.** Write the steps you would take to find out how much rain fell on one day.
2. Write about and draw what you can wear in different kinds of weather.

Music Link

Listen to "Let's Talk About the Weather" on the Science Songs CD.

LOG ON e-Review Summaries and quizzes online at www.macmillanmh.com

Lesson 2

The Water Cycle

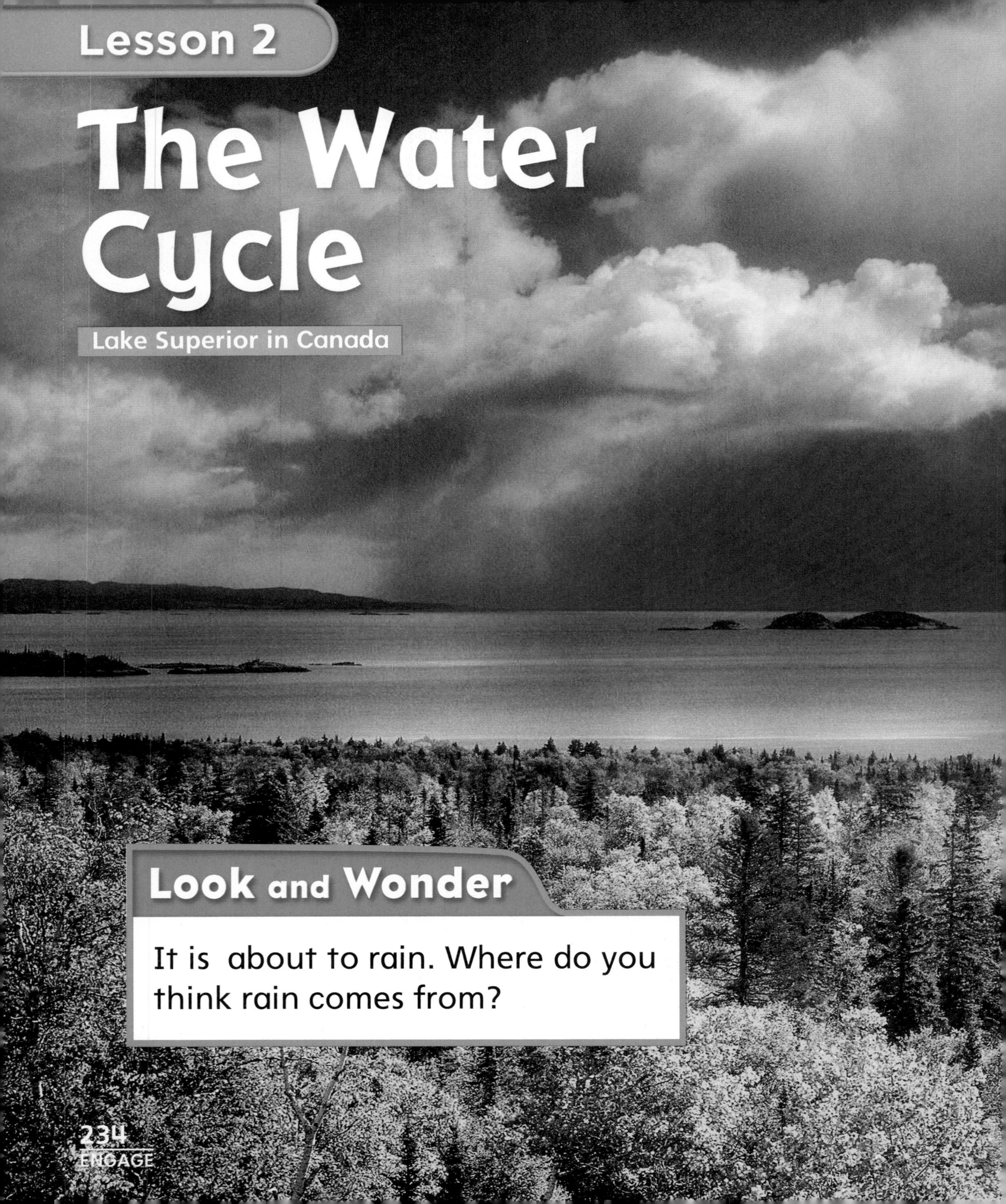

Lake Superior in Canada

Look and Wonder

It is about to rain. Where do you think rain comes from?

Explore

Inquiry Activity

What kinds of clouds can you see?

What to Do

1. **Observe.** Go outside each day for one week. Look at the clouds.

2. **Record Data.** Draw a picture of the clouds you see each day.

3. **Communicate.** Describe how the clouds change over the week.

Explore More

4. **Predict.** Do you think the clouds will look the same next week?

You need

crayons

paper

Step 2

Read Together and Learn

Vocabulary

water vapor

cloud

SCIENCE QUEST Explore the water cycle with the Junior Rangers.

What makes it rain or snow?

The Sun warms the land, air, and water. As the Sun warms water, some water turns into water vapor.

Water vapor is water that goes up into the air. It is too small to see.

The Water Cycle

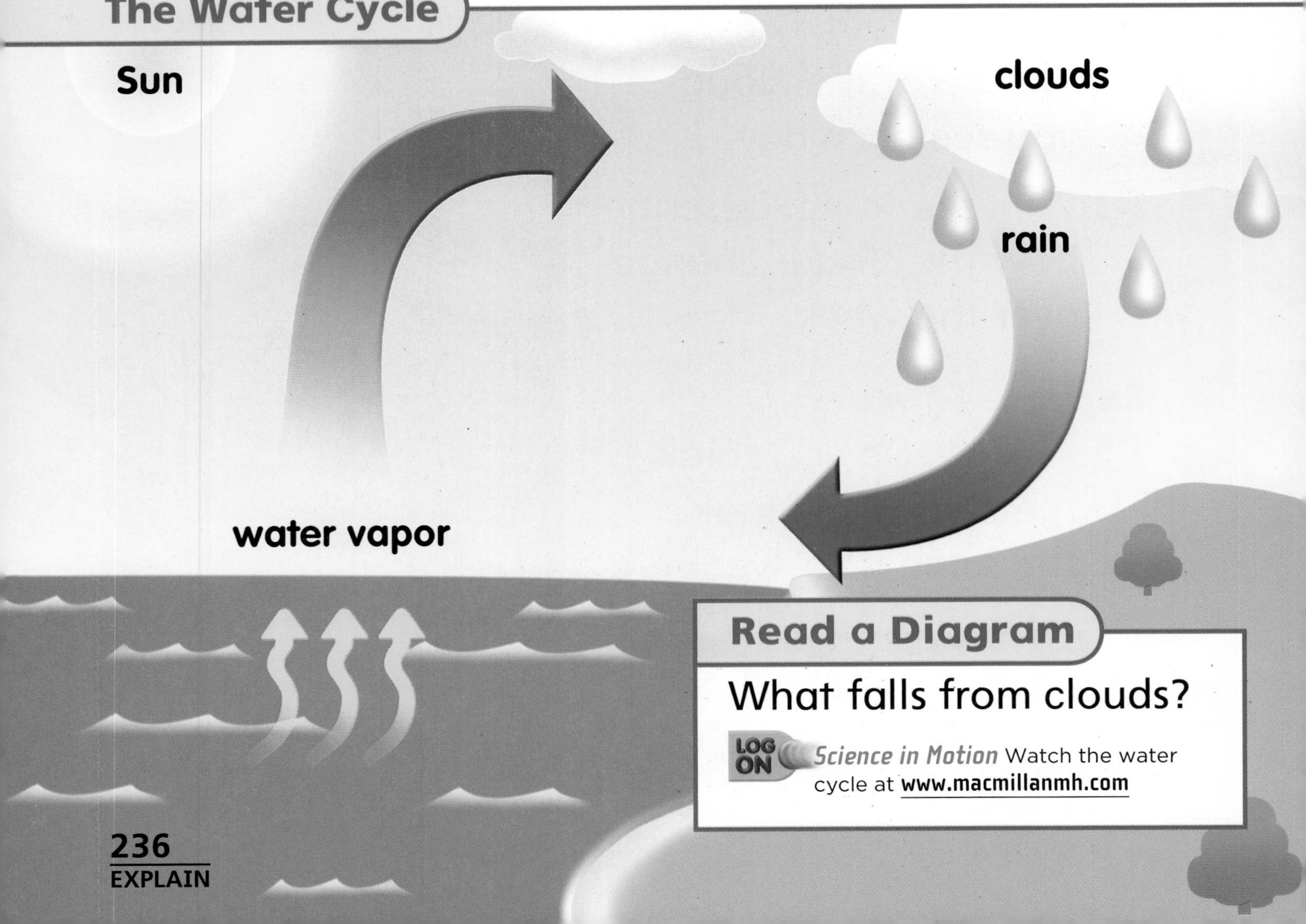

Read a Diagram

What falls from clouds?

LOG ON *Science in Motion* Watch the water cycle at www.macmillanmh.com

As water vapor cools in the sky, it turns back into drops of water or bits of ice. **Clouds** are made of both.

When the water drops or bits of ice get bigger or colder, they can fall to Earth as rain, snow, sleet, or hail.

▲ **Hail can be big or small lumps of ice.**

▲ **Rain is drops of water.**

▲ **Sleet is frozen rain.**

✔ What are clouds made of?

What are some different kinds of clouds?

Clouds do not always look the same. Some clouds look very thin. Other clouds seem big and puffy. That is because there are many different kinds of clouds.

Quick Lab

Put water in a bag and watch the water cycle.

Cumulus clouds are made mostly of water drops.

Cirrus clouds are made mostly of thin bits of ice and look feathery.

Cumulonimbus clouds are thunder clouds that bring rain and storms.

What kind of clouds could help you predict a storm?

Think, Talk, and Write

1. Classify. Name some different things that fall from clouds.
2. Write about the water cycle.

Health Link

Act out different types of weather. Show what can fall from clouds.

LOG ON e-Review Summaries and quizzes online at www.macmillanmh.com

Lesson 3

Spring and Summer

Look and Wonder

This tree has flowers. What kinds of flowers do you see in spring?

Explore

Inquiry Activity

Do seeds grow faster when it is warm or cold?

You need

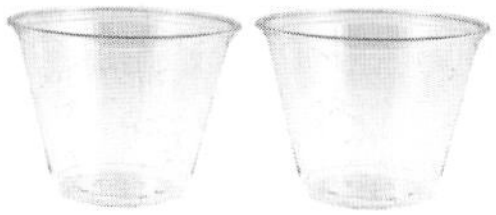

cups

soil

radish seeds

aluminum foil

What to Do

1. Plant radish seeds in two cups of soil. Cover the cups with foil.
2. Put one cup in a warm place. Put the other cup in the refrigerator.
3. **Predict.** Which do you think will grow faster?
4. **Compare.** Check the cups every day. What happens?

Explore More

5. **Infer.** What do you think will happen if you take the seeds out of the refrigerator and put them in a warm place?

Step 1

Read Together and Learn

Vocabulary

season
spring
summer

What happens in spring?

The weather changes in many places during the year. A **season** is a time of year.

Spring, summer, fall, and winter are the four seasons. In **spring**, there are many hours of sunlight.

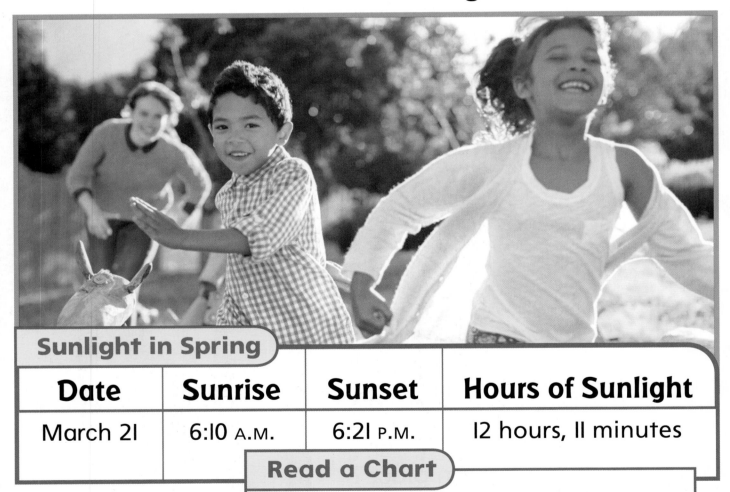

Sunlight in Spring

Date	Sunrise	Sunset	Hours of Sunlight
March 21	6:10 A.M.	6:21 P.M.	12 hours, 11 minutes

Read a Chart

If you played outside on March 21, what time would it get dark?

Sunlight warms the land, air, and water. It can also rain a lot in spring. Sunlight and rain help plants grow. Growing plants are food for many animals.

Spring weather where you live is probably similar each year.

Quick Lab

Draw what kind of clothes you wear in different seasons.

What is spring like where you live?

▼ In spring plants begin to sprout and many animals are born.

What happens in summer?

Summer is the season after spring. It is the warmest season. The weather can be very sunny and dry in summer.

Sunlight in Summer

Date	Sunrise	Sunset	Hours of Sunlight
June 21	5:43 A.M.	8:37 P.M.	14 hours, 54 minutes

▼ **In summer it may be very hot. A cold drink can help you cool off.**

Lots of sunlight helps plants grow. Many plants grow fruits.

There is a lot of food for animals to eat. Young animals grow bigger and stronger in the summer.

✔ **Is summer where you live the same every year?**

▲ **In summer there are more plants for animals to eat.**

Think, Talk, and Write

1. **Main Idea and Details.** What is a season?
2. Write about and draw what happens in spring.

Health Link

Why do you need to wear sunblock when you are outside?

LOG ON e-Review Summaries and quizzes online at www.macmillanmh.com

Museum Mail Call

What is spring like in other places? Scientists at the American Museum of Natural History collect stories to learn about people around the world.

Dear Museum,

Be Méeybaan, how are you? I live in Pakistan. My people, the Hunza, live in the mountains.

It is April. Spring is here. We plant seeds and celebrate. It does not rain much here.

Each spring, the large, icy glaciers above our village melt. We dig ditches so the water from the melted glaciers can flow onto our land. This helps our seeds grow!

From,

Nazir

AMERICAN MUSEUM OF NATURAL HISTORY

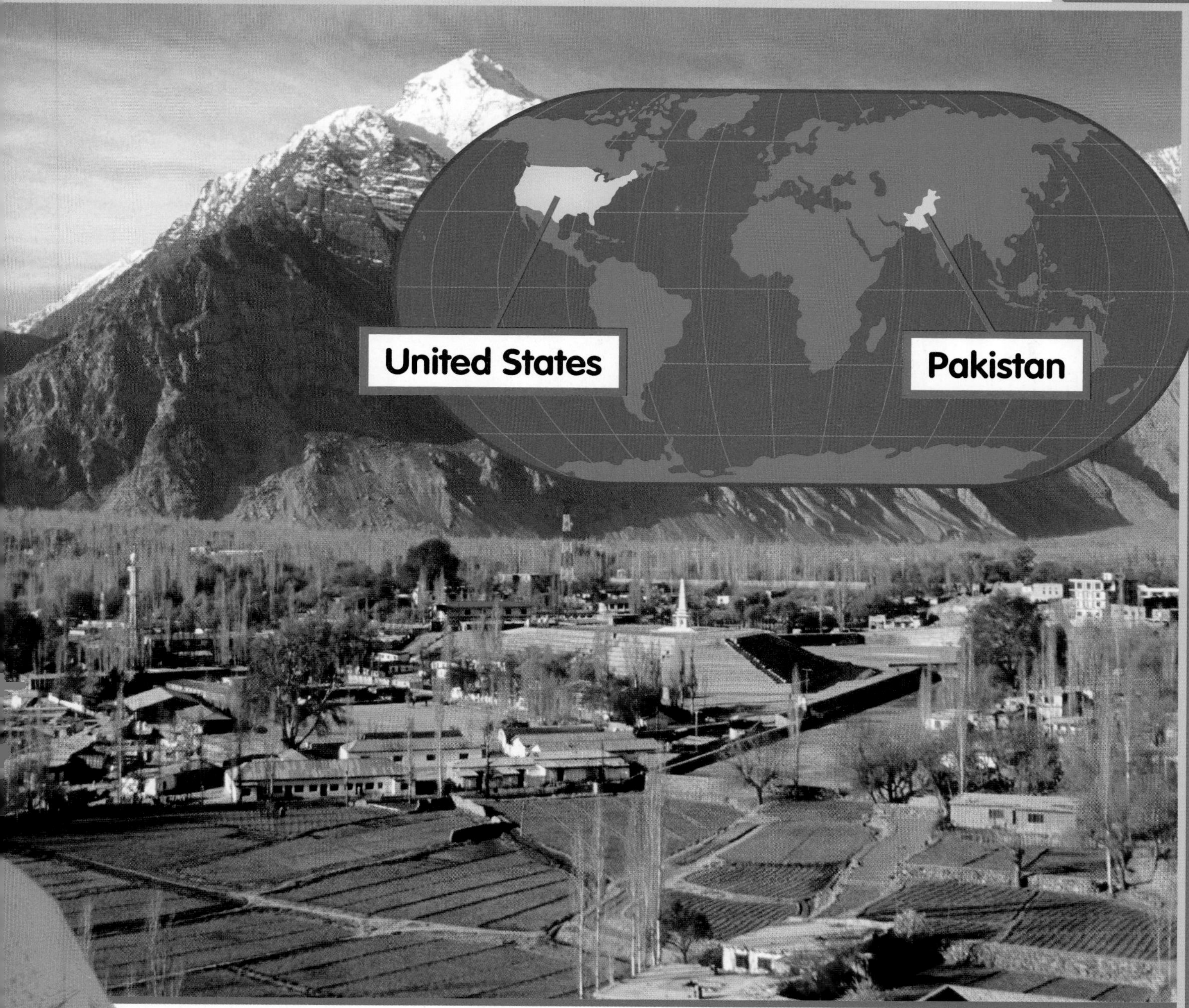

▲ The Hunza live in the mountains of Pakistan.

Talk About It

Main Idea and Details. What happens in spring that helps the Hunza farmers' seeds grow?

Lesson 4

Fall and Winter

Look and Wonder

What is the season here? How do you know?

Explore

Inquiry Activity

How do sweaters keep us warm?

What to Do

1. Fill two jars with warm water. Wrap one jar with a thick cloth.
2. **Predict.** Which jar will stay warmer? Why?
3. **Measure.** Measure the temperature of the water in each jar with a thermometer. Record your results. Measure again in 10 minutes.

Explore More

4. **Infer.** How is wrapping cloth around a jar like wearing a sweater on a cool fall day?

You need

two jars

water

cloth

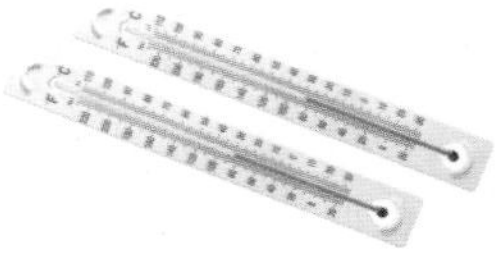

two thermometers

Step 3

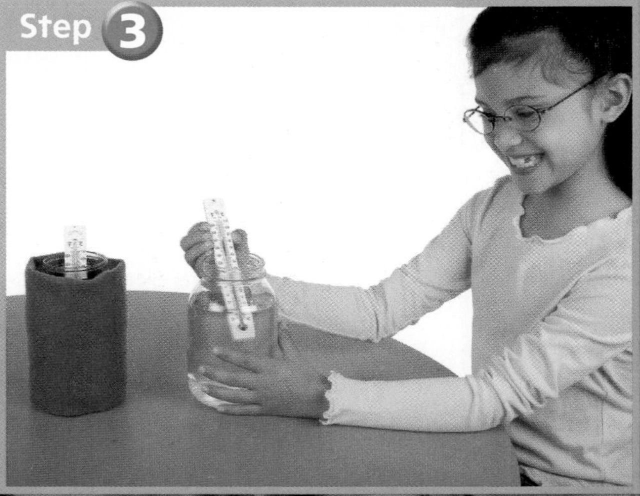

Read Together and Learn

Vocabulary

fall

winter

What happens in fall?

Fall is the season after summer. In fall, there are fewer hours of sunlight than in summer.

Less sunlight makes the air cooler. Some leaves change color and fall off trees in fall.

Sunlight in Fall

Date	Sunrise	Sunset	Hours of Sunlight
September 21	6:55 A.M.	7:07 P.M.	12 hours, 12 minutes

Many fruits get ripe in fall. People can pick the fruits and eat them.

In fall animals begin to get ready for winter. Some animals grow thicker fur to keep warm. Some even move to warmer places.

Quick Lab

Find out how leaves look different in each season.

What is fall like where you live?

These birds fly south for winter.

This chipmunk stores nuts in fall so that it has food to eat in winter.

FACT When it is fall where you are, there are other places on Earth where it is spring.

What happens in winter?

Winter is the coldest season of the year. In winter there are fewer hours of sunlight. It may even snow.

Less sunlight means some plants die. The last of the leaves fall off some trees.

Seasons

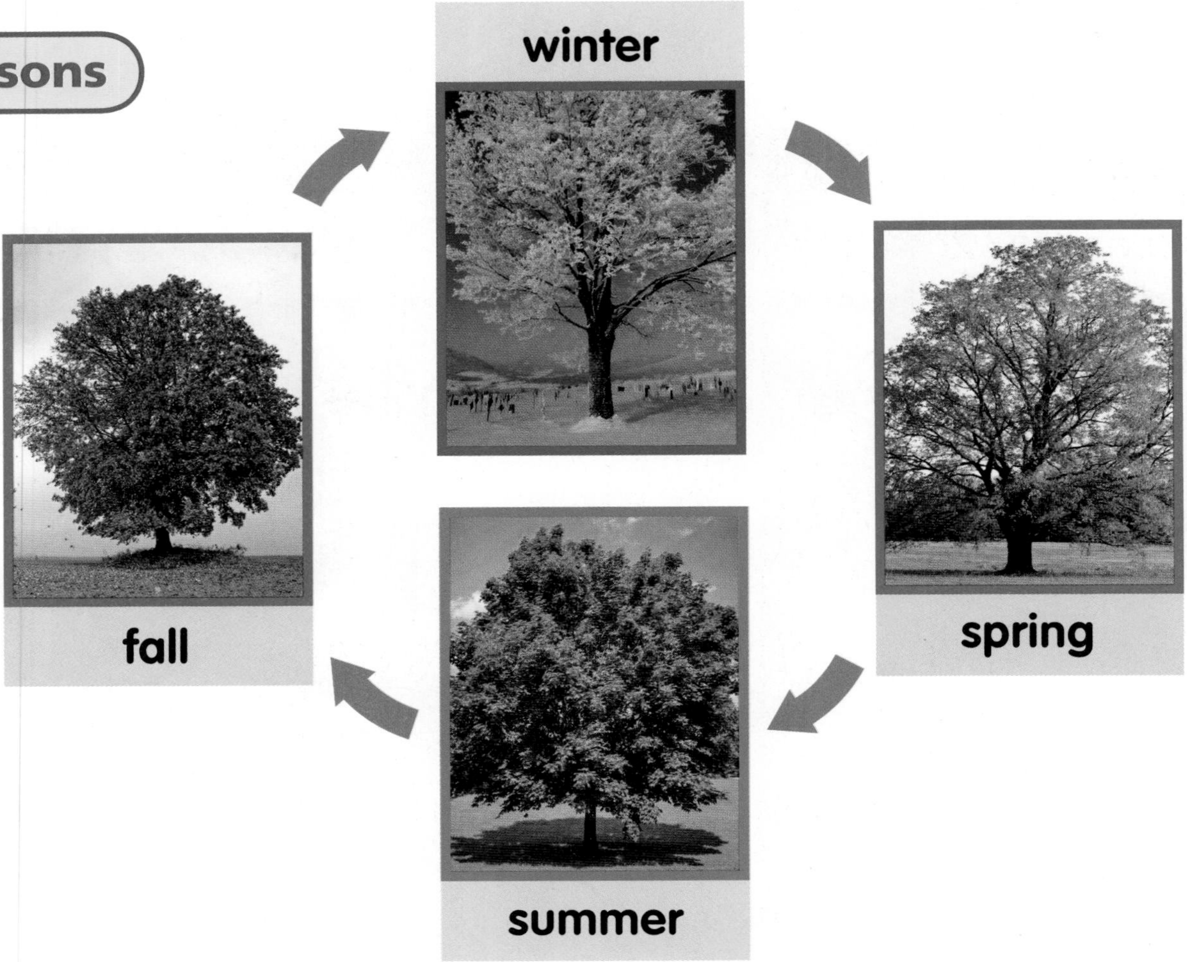

Read a Diagram

What happens to the tree in each season?

There are fewer plants in winter. There is not a lot of food for animals to eat.

Some animals search for food. Others eat food they stored in fall. Some animals go to sleep until spring.

▲ This dormouse will rest here all winter.

Sunlight in Winter

Date	Sunrise	Sunset	Hours of Sunlight
December 21	7:23 A.M.	4:49 P.M.	9 hours, 26 minutes

✓ **What is winter like where you live?**

Think, Talk, and Write

1. **Summarize.** How do animals get food in winter?
2. Write about what the weather is like in fall.

Social Studies Link

Describe different things people do in each season where you live.

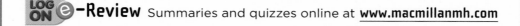

LOG ON e-Review Summaries and quizzes online at www.macmillanmh.com

Seasons Change

Look closely at the pictures below. What is the same in each picture? What is different? Which season does each picture show? How do you know?

Write About It

Write about one of the pictures. Describe the weather. What could you wear and do if you were there?

Remember

Use words to tell how something looks, feels, and sounds.

LOG ON e-Journal Write about it online at www.macmillanmh.com

Weather Graph

Anna asked her friends which activity they liked to do best on sunny days. She made a bar graph to show what she found out.

Sunny Day Activities

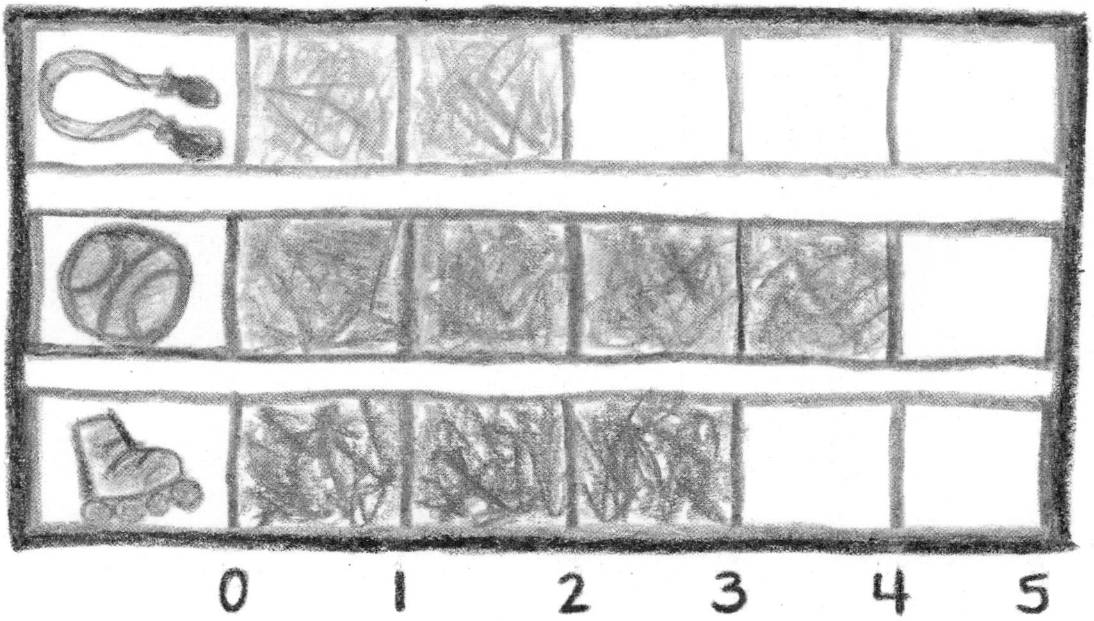

Make a Graph

Ask your classmates if they like to read a book, play a game, or draw a picture on a rainy day. Make a bar graph to show what you find out.

Remember

A bar graph needs a title.

I Read to Review

My Seasons Book

Spring can be wet.

I need my rain boots.

I like all of the new plants.

Summer can be sunny.

I need my hat.

I like to play in the sand.

Fall can be cool.
I need my jacket.
I like to play in the leaves.

Winter can be cold.

I need my mittens.

I like to play in the snow.

CHAPTER 7 Review

Vocabulary

Use each word once to complete the sentences.

cloud
summer
thermometer
winter

1. The warmest season is _____.

2. A _____ is made of tiny droplets of water or small bits of ice.

3. The season with the fewest hours of daylight is _____.

4. We can measure the temperature of the air with a _____.

Answer the questions below.

5. How is the weather different in the pictures below?

6. **Predict.** What do you think will happen to the puddle in this picture?

7. **Main Idea and Details.** In which season do trees lose leaves and animals move to warmer places?

The Big Idea

8. What do you know about weather?

CHAPTER 8

The Sky

The Big Idea

What can you see in the sky?

Key Vocabulary

star an object in the sky that makes its own light (page 266)

Sun the star closest to Earth (page 267)

Moon a ball of rock that moves around Earth (page 280)

planet a very large object that moves around the Sun (page 282)

More Vocabulary

rotate, page 272

phases, page 281

Lesson 1

The Sky Above

Look and Wonder

It is night. What kinds of things can you see in this night sky?

Explore

Inquiry Activity

What can you see in the sky?

What to Do

1. **Observe.** Look at the sky during the day. Then look at a picture of the sky at night.
   △ **Be Careful.** Do not look directly at the Sun.

2. **Record Data.** Make a list of what you see in the day sky and the night sky. Do you see the same things?

Explore More

3. **Infer.** What do you think happens to stars during the day?

Read Together and Learn

Vocabulary

star

Sun

SCIENCE QUEST Explore the Sun with the Junior Rangers

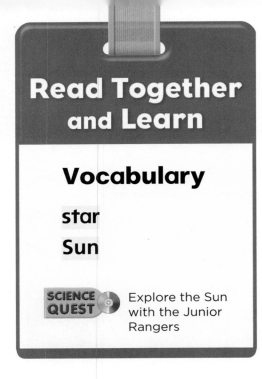

What is in the sky?

At night you might see clouds, the Moon, or stars. **Stars** are objects in the sky that make their own light.

Stars can make patterns in the sky. Stars look tiny because they are far away.

▼ **Telescopes can help you observe faraway things like stars.**

The **Sun** is the star closest to Earth. It lights the sky and Earth during the day.

In the daytime, sunlight is so bright that you can not see other stars. But they are still there.

Quick Lab

Compare the different things that you can see in the sky.

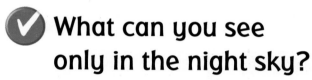

✓ **What can you see only in the night sky?**

There are more stars than you can count.

FACT Sometimes you can see the Moon during the day.

Why is the Sun important?

The Sun and other stars make energy in the form of heat and light.

When the Sun looks low in the sky, the temperature can feel cool. When the Sun looks high in the sky, it can feel warmer.

Sun's Warmth

Read a Photo

In which picture do you think it is warmer? Why?

The Sun's energy warms the air, land, and water. Without the Sun, it would be too cold for us to live on Earth.

If there were no Sun, Earth would always be dark. Without light, we would not be able to see.

✓ **Why is the Sun important?**

Think, Talk, and Write

1. **Problem and Solution.** How can you see faraway objects in the sky?
2. Write a list of what you can see in the day and night skies.

Music Link

Listen together to the Science Song "A Sun for All Seasons."

LOG ON **e-Review** Summaries and quizzes online at www.macmillanmh.com

Lesson 2

Earth Moves

Look and Wonder

When something blocks the Sun's light, it makes a shadow. Where do you think the Sun is here?

Explore

Inquiry Activity

What causes shadows?

What to Do

1. **Observe.** Put a craft stick in clay on a piece of paper. Shine a flashlight on the craft stick.
2. **Record Data.** Trace the shadow and label it. Move the flashlight to another spot. Trace and label the new shadow.
3. **Communicate.** How did the shadow change?

Explore More

4. **Investigate.** How can you make a short shadow? How can you make a long shadow?

You need

craft stick

modeling clay

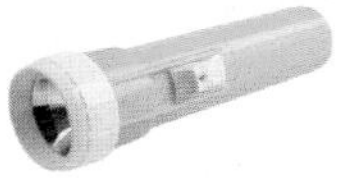

flashlight

paper

pencil

Step 1

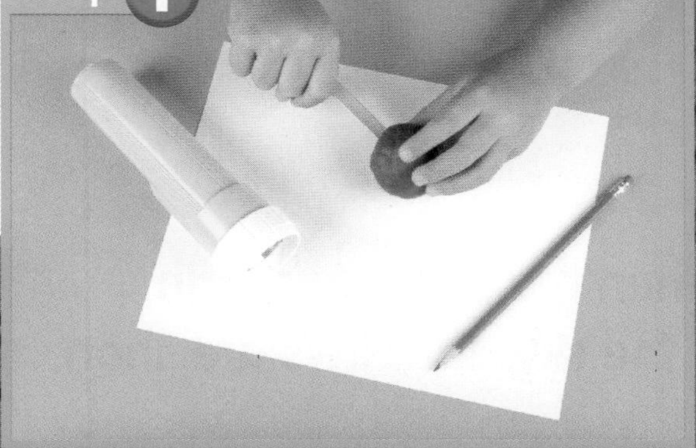

Read Together and Learn

Vocabulary

rotate

What causes day and night?

Have you seen the Sun rise and set in the sky? The Sun is not really moving. Earth is.

We can not feel it, but Earth rotates. To **rotate** is to spin very fast. You can tell Earth moves by looking at shadows.

▲ When the Sun looks high in the sky, shadows are short.

▲ When the Sun looks low in the sky, shadows are long.

Earth rotates, making night and day. When the place you are on Earth faces the Sun, the sky is light. It is day.

When Earth rotates you away from the Sun, the sky is dark. It is night.

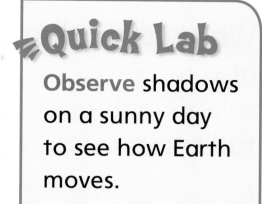

Quick Lab

Observe shadows on a sunny day to see how Earth moves.

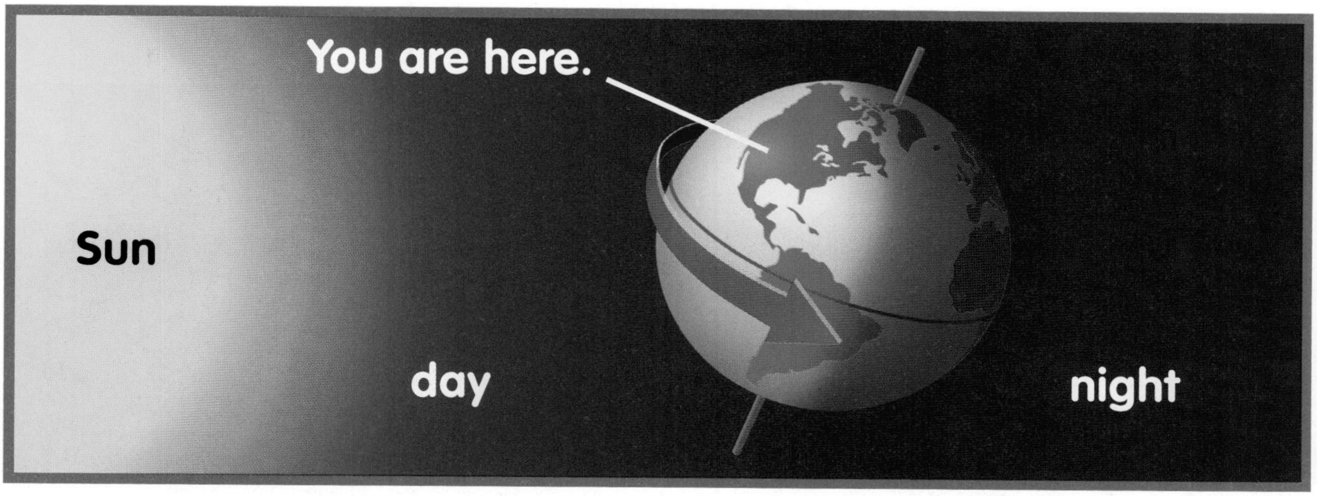

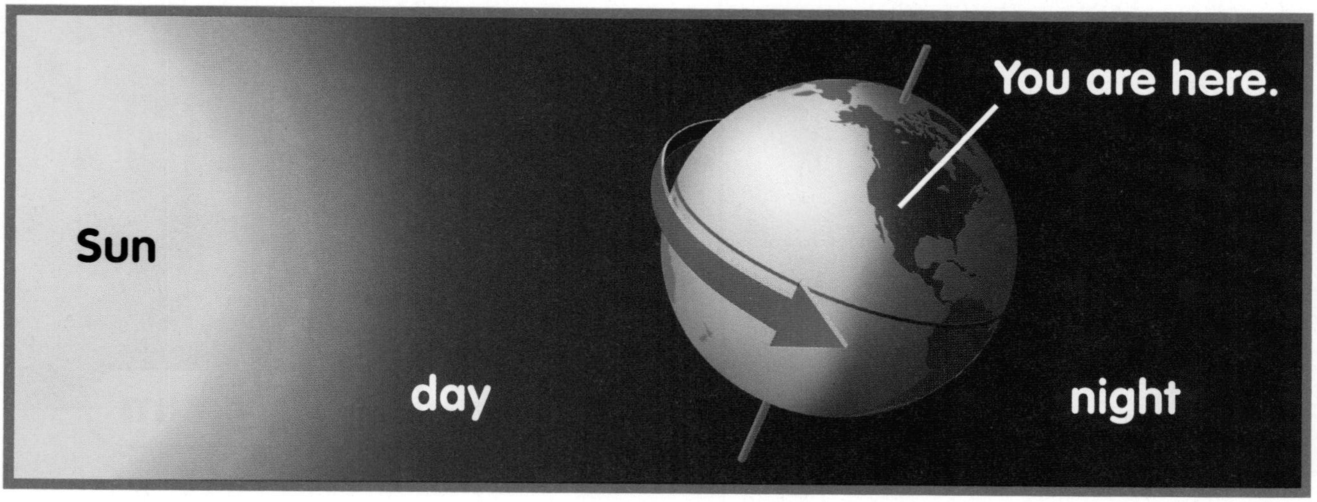

✔ What causes day and night?

What causes a year?

While Earth rotates each day, it also moves around the Sun. It takes Earth one year to make a full trip around the Sun.

As Earth moves, part of it leans toward the Sun. The other part leans away from the Sun.

One Full Year

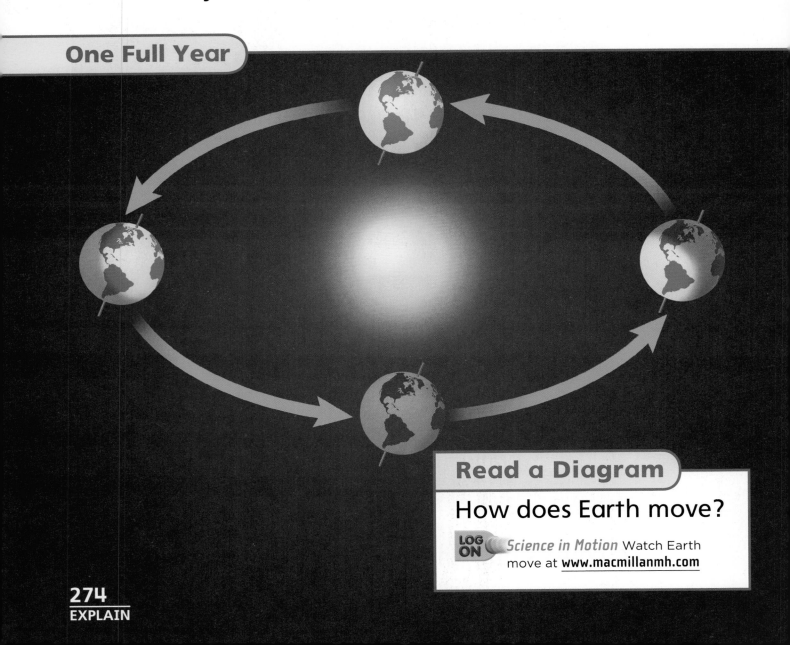

Read a Diagram

How does Earth move?

LOG ON *Science in Motion* Watch Earth move at **www.macmillanmh.com**

This makes the four seasons.

✔ How can you tell that Earth moves?

Think, Talk, and Write

1. **Compare and Contrast.** What are two ways that Earth moves?
2. Write about and draw how shadows can look at different times of day.

Art Link

Use clay to make models of Earth and the Sun. Show how Earth moves.

LOG ON e-Review Summaries and quizzes online at www.macmillanmh.com

Time of Day

The position of the Sun seems to change during the day. This is because Earth rotates, or spins. The sky can tell us a lot about the time of day.

Write About It

Write about the time of day you think this picture was taken. How do you know? What do you think the temperature feels like here?

Remember
Use words that describe the Sun.

LOG ON e-Journal Write about it online at www.macmillanmh.com

Measure Time

Ricky estimated how long it would take him to do different activities. Then he used a clock to find out how long they really took. He made a chart.

How long does it take?

Activity	Estimated time	Actual time
Brushing teeth	1 minute	2 minutes
Reading a book	20 minutes	30 minutes
Sleeping	16 minutes	540 minutes

Estimate Activities

Make a chart like Ricky's. Do your activities take longer than you estimated? How can you find out?

Remember

You estimate when you guess how much time it takes to do something.

Lesson 3

Earth's Neighbors

Look and Wonder

We see the Moon because the Sun's light shines on it. What do you notice about the Moon here?

Explore Inquiry Activity

Does the Moon always look the same?

What to Do

1. **Record Data.** Go outside with an adult at night. Look at the Moon. Draw a picture of what it looks like.
2. **Compare.** Take your picture to school. Compare it with the pictures below.

3. **Communicate.** Does your Moon drawing look like one of these? How is it the same? How is it different? Share your drawing with a classmate.

Explore More

4. **Investigate.** Look at and draw the Moon each night for one month. How does it change? Do you see a pattern?

Read Together and Learn

Vocabulary

Moon
phase
planet

How does the Moon look?

The Moon rises and sets in the sky. The **Moon** is a ball of rock that moves around Earth. It takes about a month for the Moon to go once around Earth.

The Moon does not make its own light. It looks bright because the Sun's light shines on it.

▼ **These holes on the Moon are called craters.**

As the Moon moves around Earth, the lit part of the Moon that we see changes. This is why the Moon seems to have different shapes.

Quick Lab

Make a model of the Moon's craters.

The different Moon shapes we see are called **phases**. Each month you see the same Moon phases.

Moon Phases

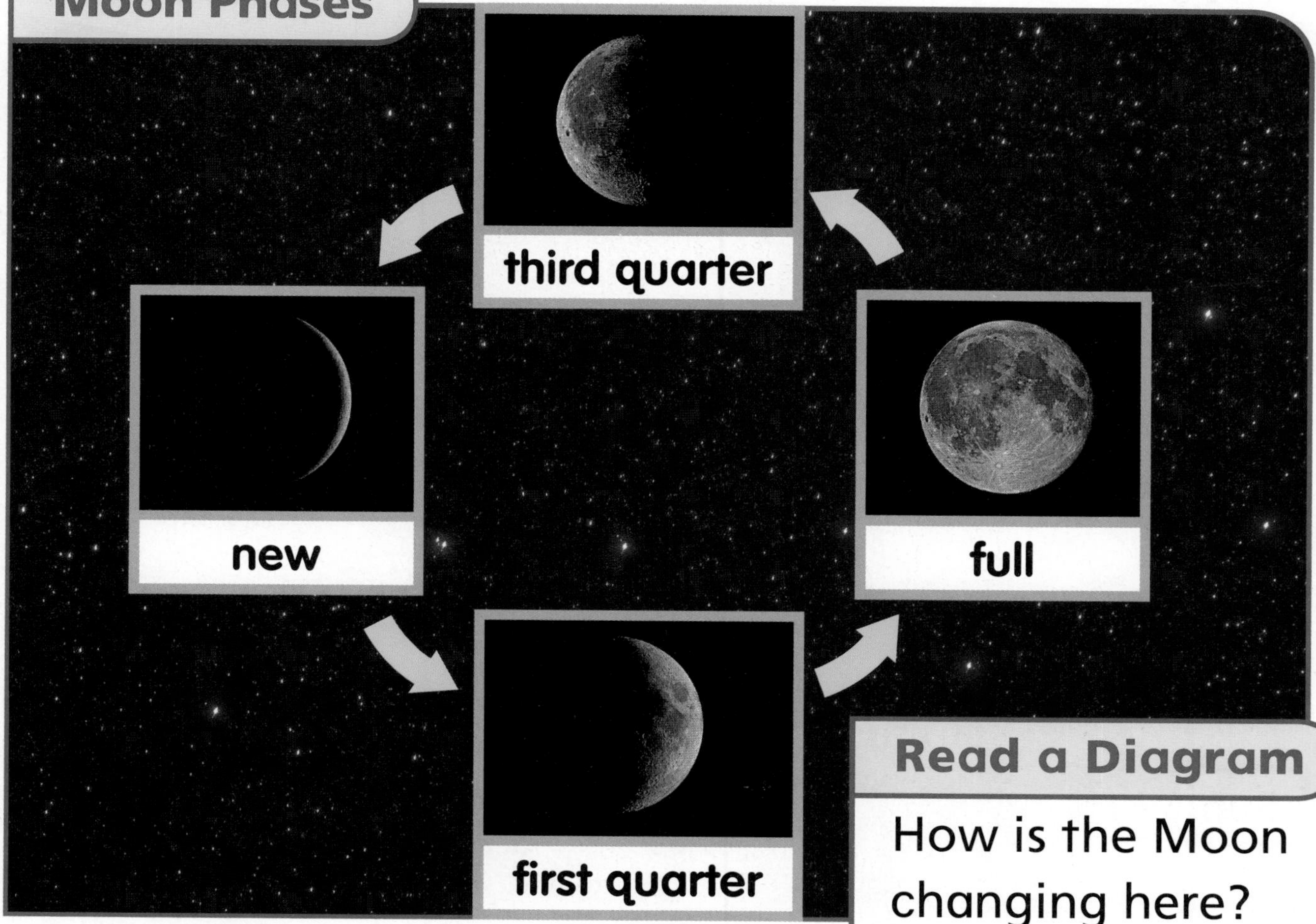

Read a Diagram

How is the Moon changing here?

✔ Why does the Moon look bright?

What are planets?

Have you ever seen a light in the night sky that looks bigger or brighter than the other stars? It might be a planet.

A **planet** is a very large object that moves around the Sun. There are eight planets that move around the Sun. Earth is one planet.

These are the eight planets that move around the Sun.

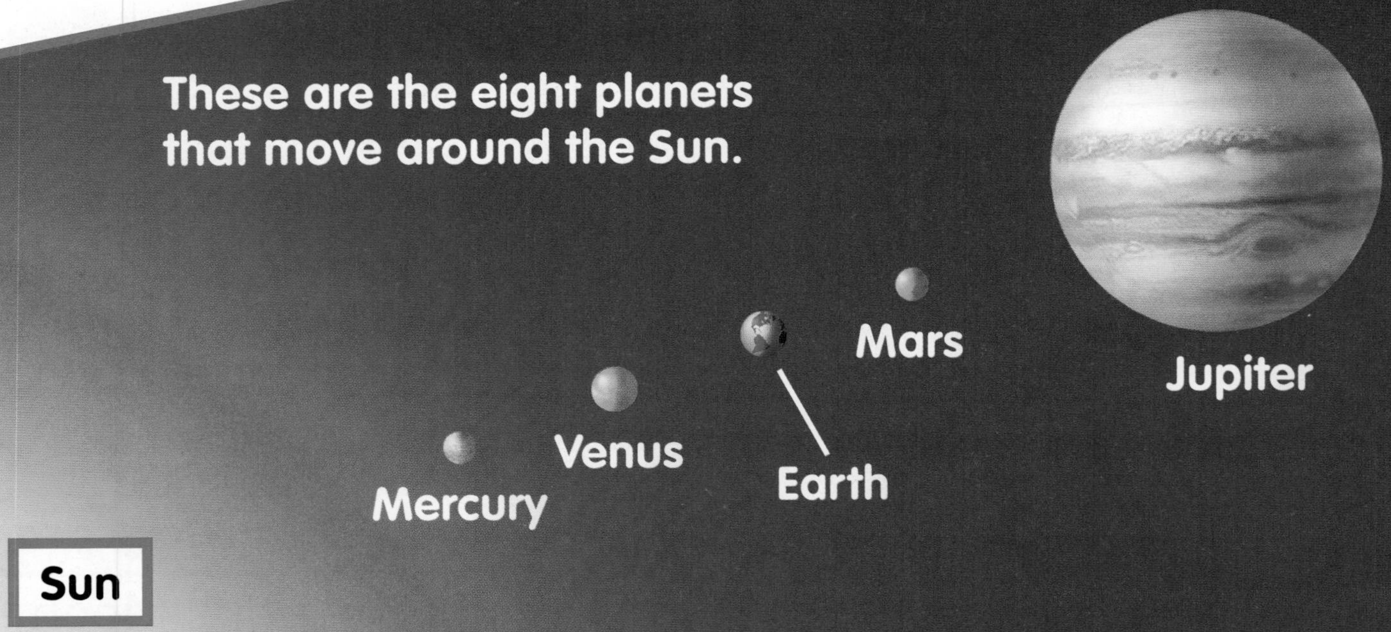

FACT None of the other planets around our Sun has humans, plants, or animals like we do on Earth.

The other planets do not look like Earth. Some are smaller than Earth. Others are larger. The planets closer to the Sun are warmer. The planets farther away are cooler.

✓ **What is a planet?**

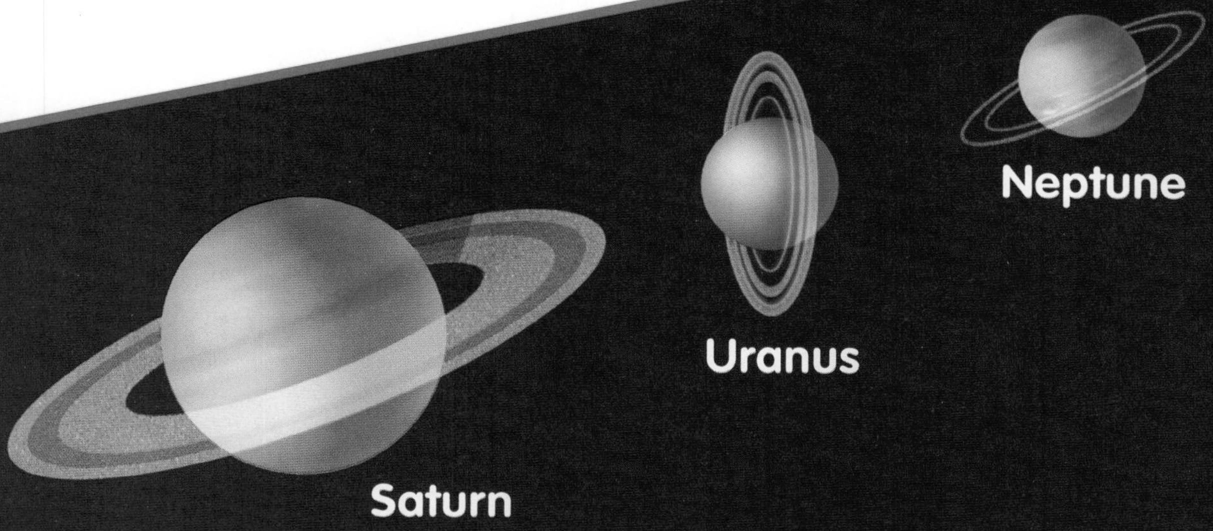

Think, Talk, and Write

1. **Cause and Effect.** Why do we see different Moon shapes?
2. Write about and draw the planets.

Art Link

Make a Moon journal. Draw the Moon phases over a month.

LOG ON **e-Review** Summaries and quizzes online at www.macmillanmh.com

Meet Ben Oppenheimer

Ben Oppenheimer is an astrophysicist. He studies planets. You can see some planets by looking up into the sky.

Venus and Mars are planets that shine in the night sky. Other planets are too far away to see with just your eyes. Scientists, like Ben, use telescopes to see them.

Telescopes are tools that help us see things that are far away. Ben uses them to learn more about the planets.

Ben Oppenheimer

Some telescopes stay up in space. The Spitzer Space Telescope helps scientists find planets.

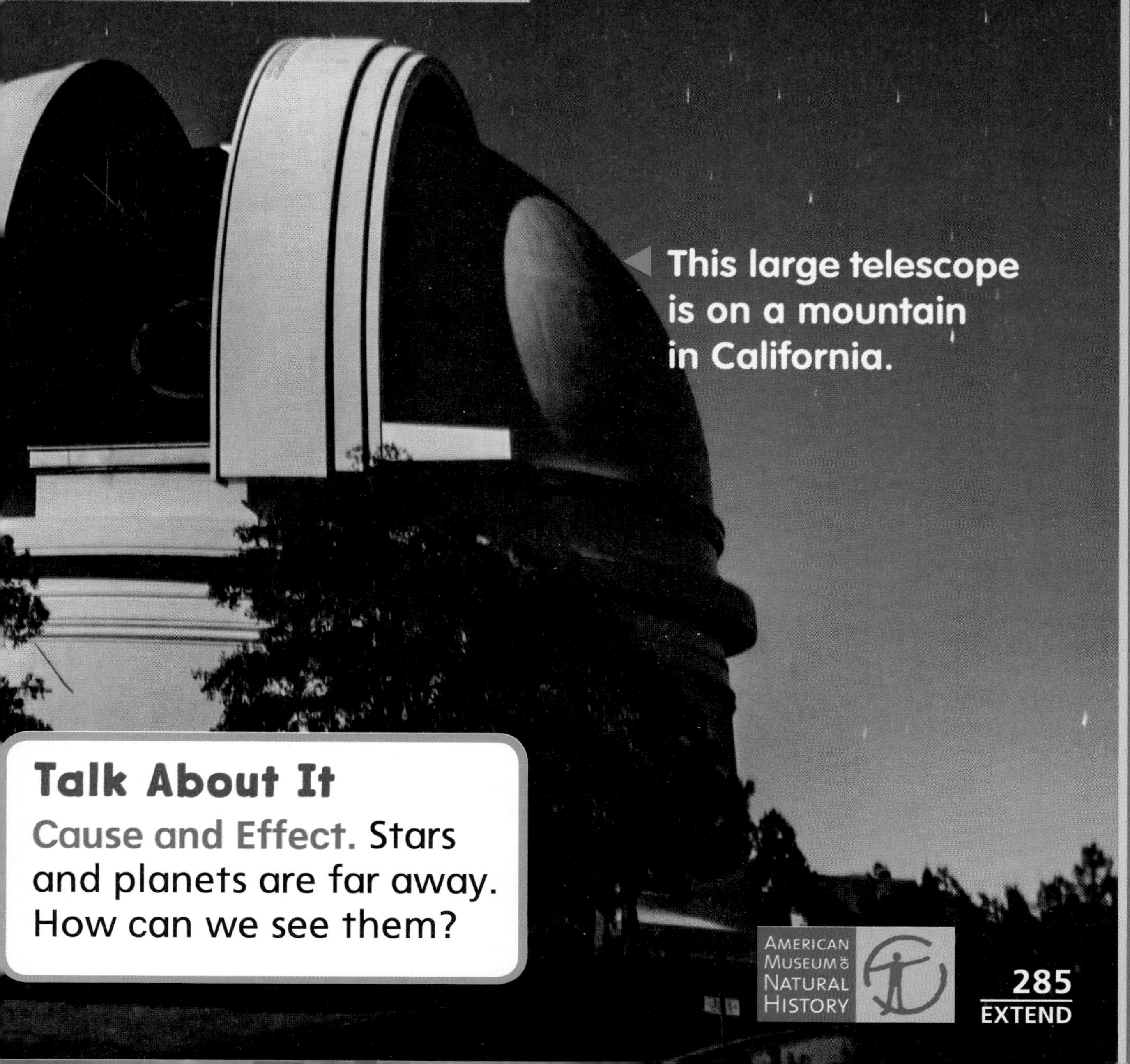

This large telescope is on a mountain in California.

Talk About It

Cause and Effect. Stars and planets are far away. How can we see them?

American Museum of Natural History

Where is the Sun?

The Sun rises in the morning.

It sets at night, when day is done.

Look down!

Shadows move and change as the Sun seems to move in the sky.

Look up!

Stars shine and the Moon moves.

We see them in the night sky.

Where is Earth?

Night and day, we are moving around the Sun.

CHAPTER 8 Review

Vocabulary

Use each word once to complete the sentences.

Moon
planets
rotates
Sun

1. The star closest to Earth is the ____.

2. The eight large objects that move around the Sun are called ____.

3. A ball of rock that moves around Earth is the ____.

4. The diagram below shows how Earth spins, or ____, making night and day.

Answer the questions below.

5. What can shadows tell us about Earth?

6. Why is the Sun important to Earth?

7. **Record Data.** Write what you can see in the day sky and night sky.

8. **Cause and Effect.** What causes a year?

9. Draw which Moon phase you think comes next.

The Big Idea

10. What can you see in the sky?

Storm Chaser

Do you like watching the weather? You could become a storm chaser. They try to get close to storms to learn how they work.

Storm chasers study weather. They use weather tools when studying a storm. Sometimes sirens warn people about a storm so they can find a safe place to stay until it is over.

storm chaser

More Careers to Think About

weather reporter

coast guard

LOG ON e-Careers at www.macmillanmh.com

Matter

An icicle forms when dripping water freezes.

Literature

National Wildlife Federation
Ranger Rick

Magazine Article

Water can be a solid, a liquid, or a gas.

from *Ranger Rick*

Where in the WORLD Is WATER?

Did you know that water covers most of Earth?

Solid Water

Water is solid when it is frozen. You can cut it into different shapes. It can fall from the sky as snow.

ice sculpture

Liquid Water

Liquid water fills rivers, lakes, and oceans. It falls from the sky as rain. It feels wet.

lake

Water as a Gas

When water boils, it changes from a liquid to a gas. The gas goes into the air.

boiling water

Talk About It

How do you use different kinds of water every day?

CHAPTER 9

Matter Everywhere

The Big Idea

What are things made of?

Key Vocabulary

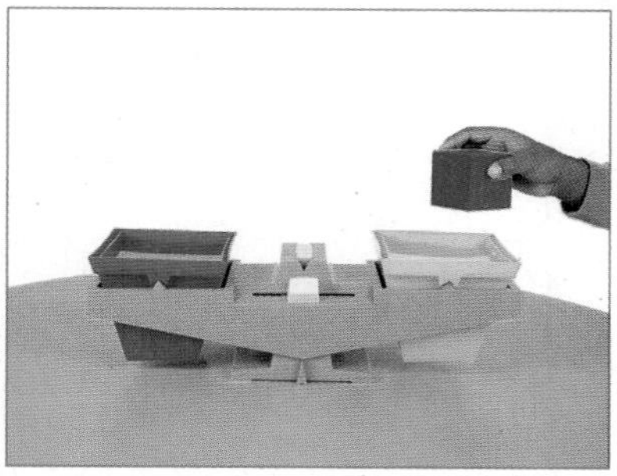

balance a tool used to measure mass (page 302)

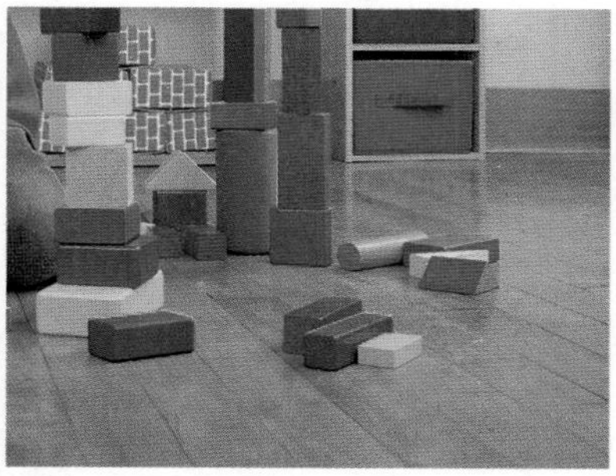

solid a state of matter that has a shape of its own (page 308)

liquid a state of matter that flows and takes the shape of its container (page 316)

gas a state of matter that does not have its own shape (page 318)

More Vocabulary

property, page 300

matter, page 301

mass, page 302

Lesson 1

Describing Matter

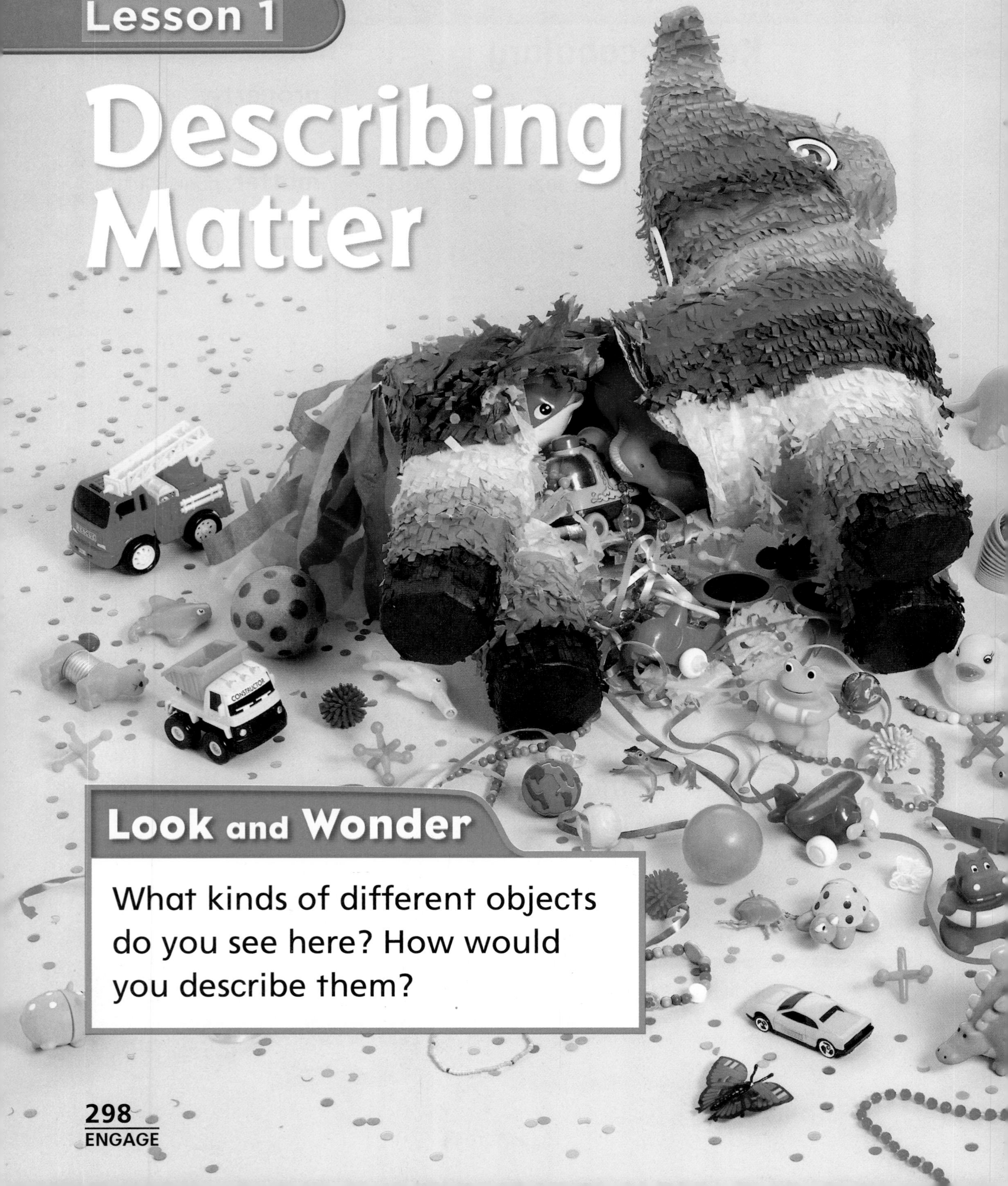

Look and Wonder

What kinds of different objects do you see here? How would you describe them?

Explore

Inquiry Activity

What can you observe about some objects?

What to Do

1. **Observe.** Look at and feel a balloon, water, and a block. Record what each one looks and feels like.
2. **Communicate.** Describe to a classmate what you observed.
3. **Compare.** How are the objects alike? How are they different?

Explore More

4. **Measure.** Find ways to measure each object. Can you measure them in the same way?

You need

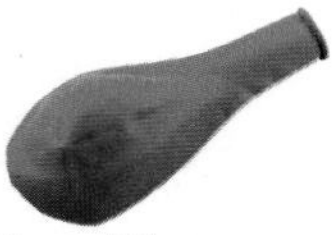

balloon

cup of water

block

Step 1

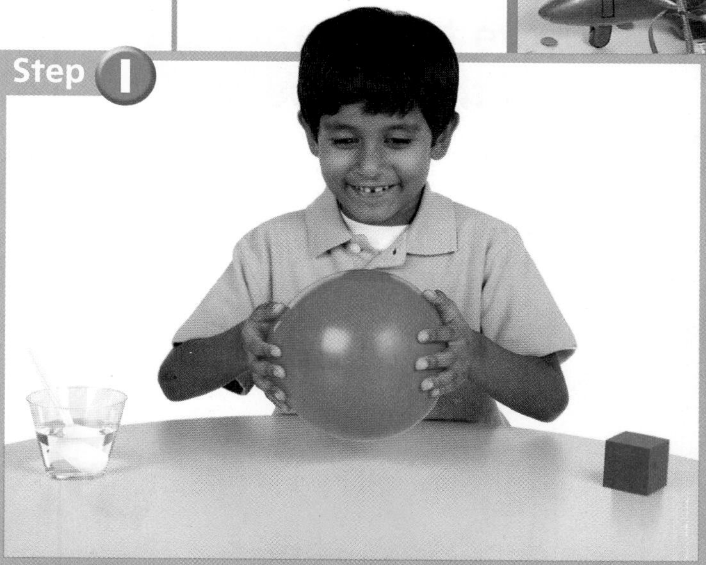

Read Together and Learn

Vocabulary

property
matter
mass
balance

▲ **Brown and soft are two properties of this toy bear.**

What is matter?

When you describe something, you talk about its properties.

Properties are how something looks, feels, smells, tastes, and sounds. Color, size, and shape are properties.

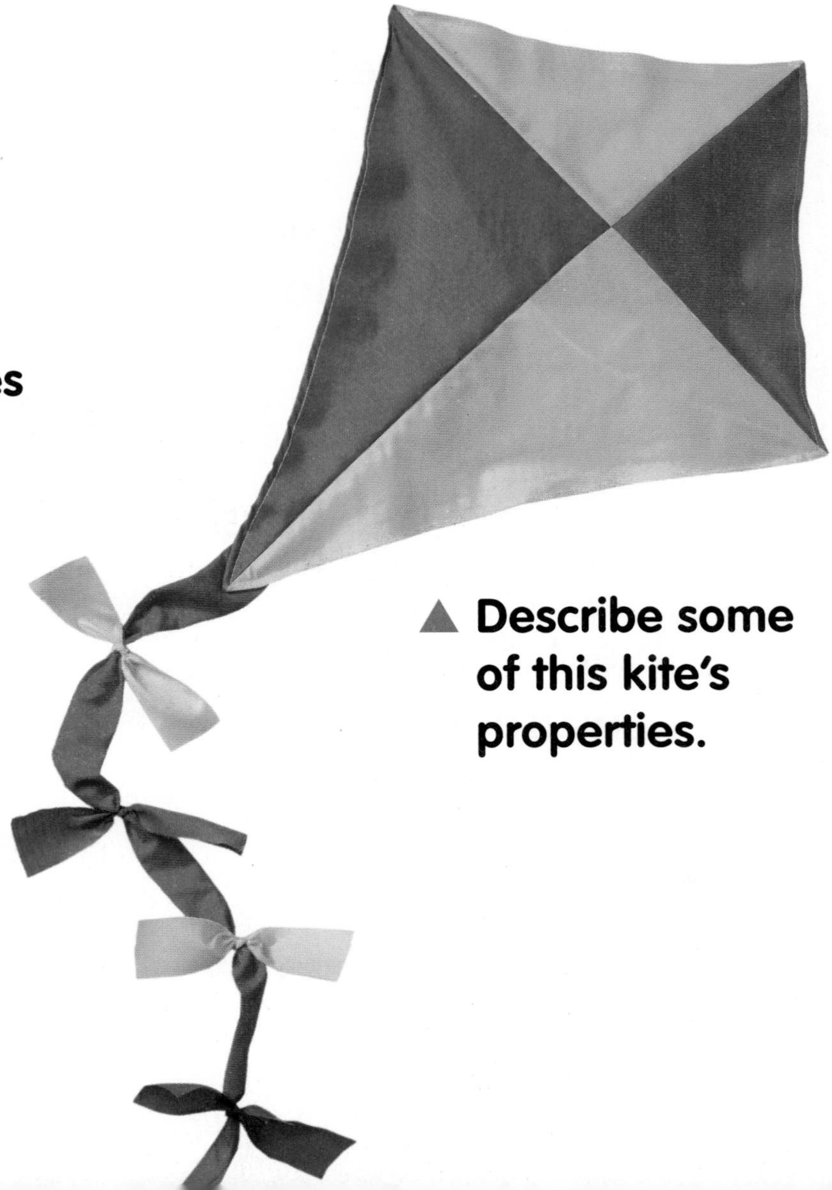

▲ **Describe some of this kite's properties.**

Every kind of matter has its own properties. **Matter** is what all things are made of.

Solids, liquids, and gases are three forms of matter. All matter takes up space.

Quick Lab

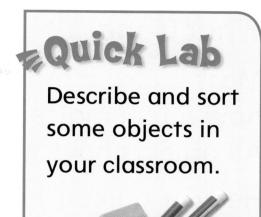

Describe and sort some objects in your classroom.

✓ **What is matter?**

▼ **The water, raft, girl, and air are all matter.**

What is mass?

Mass is also a property. **Mass** is how much matter is in an object. All objects have mass.

Heavier objects have more mass than lighter objects. A **balance** can be used to measure mass.

▼ **The metal bird has more mass than the sponge bird.**

FACT A balance does not measure weight.

Comparing Mass

Read a Photo

Which car has more mass? How could you find out?

✓ How can you measure mass?

Think, Talk, and Write

1. **Put Things in Order.** Use a balance to put three objects in order from the least mass to the most mass.
2. Write about and draw a picture of your shoes. Describe their properties.

Music Link

Listen to "What Matters Is Matter" on the Science Songs CD.

LOG ON e-Review Summaries and quizzes online at www.macmillanmh.com

A Shoe Story

Look at Manny's shoes. Where do you think they could have been?

Write About It

Write a story about Manny's shoes and where they have been. Describe the properties of the shoes.

Remember

A story has a clear beginning, middle, and end.

LOG ON e-Journal Write about it online at www.macmillanmh.com

Weigh It

A scale measures weight. Look at the scales below. Put the fruits in order from the lightest fruit to the heaviest fruit.

Weigh Yourself

Estimate your weight. Then use a scale to measure it. Check to see if you are right.

Remember

Tools improve the accuracy of estimates.

Lesson 2

Solids

Look and Wonder

Every kind of matter has its own properties. How would you describe the properties of these beads?

Explore

Inquiry Activity

How can you compare some solids?

What to Do

1. Collect five solid objects around your classroom.
2. **Compare.** Describe the objects' properties. How are they alike? How are they different? Sort them by their properties.
3. **Measure.** Use a balance to put the objects in order from the most mass to the least mass.

Explore More

4. **Classify.** What other properties can you use to sort the objects?

You need

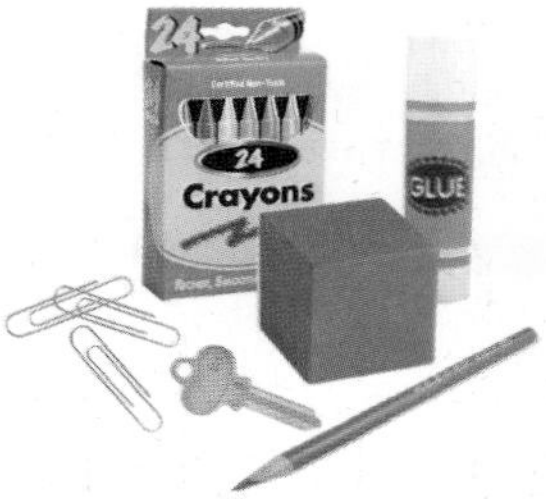

classroom objects

balance

Step 3

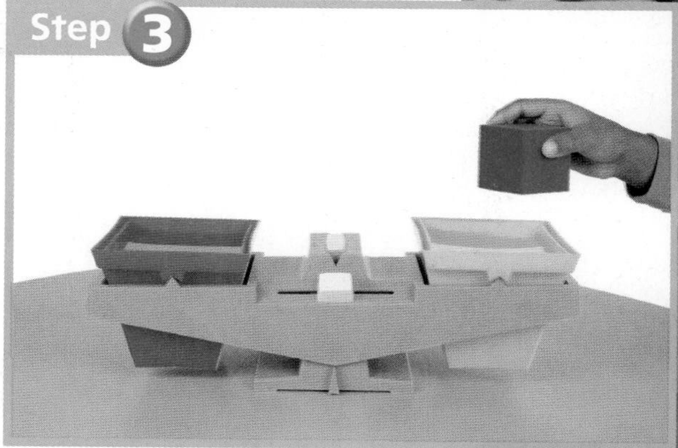

Read Together and Learn

Vocabulary

solid

SCIENCE QUEST Explore the properties of solids with the Junior Rangers.

What is a solid?

A **solid** is a form of matter. Only a solid has a shape of its own.

A solid can keep its shape even if it is moved.

Read a Photo

Describe the solids in this picture.

Building Blocks

The amount of matter in a solid always stays the same.

If you take apart a puzzle, the total amount of matter in the puzzle does not change.

The puzzle pieces and the completed puzzle have the same amount of matter.

✔ What are some solids? How do you know they are solids?

What are some properties of solids?

Solids can have many different properties.

Solids can be large or small. They come in many different shapes and colors. You can bend and fold some solids.

Solids might feel rough or smooth. How an object feels is its texture.

Solids can also be long or short. They can be wide or thin. You can use a ruler to measure some solids.

Quick Lab

Use a ruler to **measure** some solids.

What are some properties of the solids on these pages?

Think, Talk, and Write

1. **Classify.** Sort different solids by their properties.
2. Write a list of solids that you can bend.

Art Link

Use solids with different textures to make a collage. What does it feel like?

LOG ON e-Review Summaries and quizzes online at www.macmillanmh.com

Reading in Science

Real World

BUILDING BLOCKS

Do you know the story about the three little pigs? Each pig built a house from a different material to hide from the wolf.

The first pig used straw to build a house. Straw is dry, hollow grass stalks. Straw can make walls and a roof.

The second pig used wood to build a house. Wood comes from the trunks and branches of trees.

Wood is stronger than straw. A wood house can last for more than a hundred years.

The third pig used bricks to build a house. Bricks are made from hard clay.

Bricks are very strong. A brick house can last for more than a thousand years.

Talk About It

Predict. Which one of these materials would make the strongest building? Why?

AMERICAN MUSEUM OF NATURAL HISTORY

Lesson 3

Liquids and Gases

Look and Wonder

This boy is swimming in water. Why do you think there are bubbles in the water?

Explore

Inquiry Activity

What are some properties of a liquid?

What to Do

1. **Measure.** Fill a dropper with colored water. Place drops of water next to each other on wax paper.
2. **Observe.** Use a toothpick to move the drops. What happens to the drops?
3. **Communicate.** What are some properties of water?

Explore More

4. **Infer.** Do liquids have their own shape? How do you know?

You need

dropper

colored water

wax paper

toothpicks

Step 1

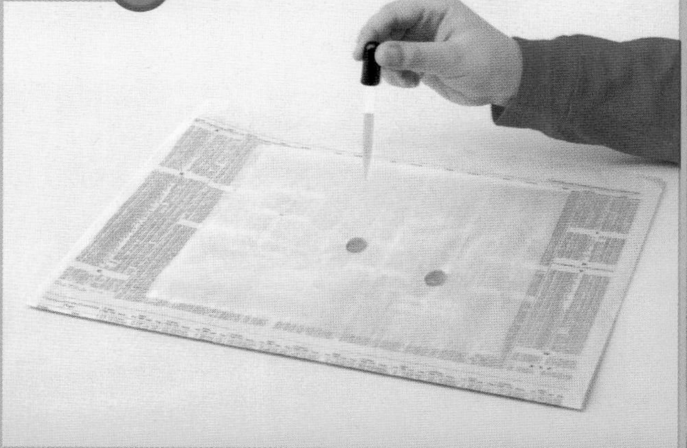

Read Together and Learn

Vocabulary

liquid

gas

SCIENCE QUEST Explore the properties of liquids and gases with the Junior Rangers.

What are some properties of liquids?

A **liquid** is a form of matter. Like solids, liquids have mass and take up space.

Liquids do not have a shape of their own. They take the shape of whatever they are in. Liquids flow when you pour them.

Liquids like honey and ketchup flow slowly.

Liquids like milk and oil flow quickly.

You can use a measuring cup to measure liquids. A measuring cup measures how much space a liquid takes up.

✔ How are liquids like solids? How are they different?

▼ **How is the liquid in the girl's glass changing shape?**

▲ **The amount of liquid in these containers is the same.**

What are some properties of gases?

A **gas** is a form of matter, too. Like liquids, gases do not have a shape of their own.

Gases spread to fill all the space of whatever they are in. Gases spread evenly in the space they are in.

Gas Changes Shape

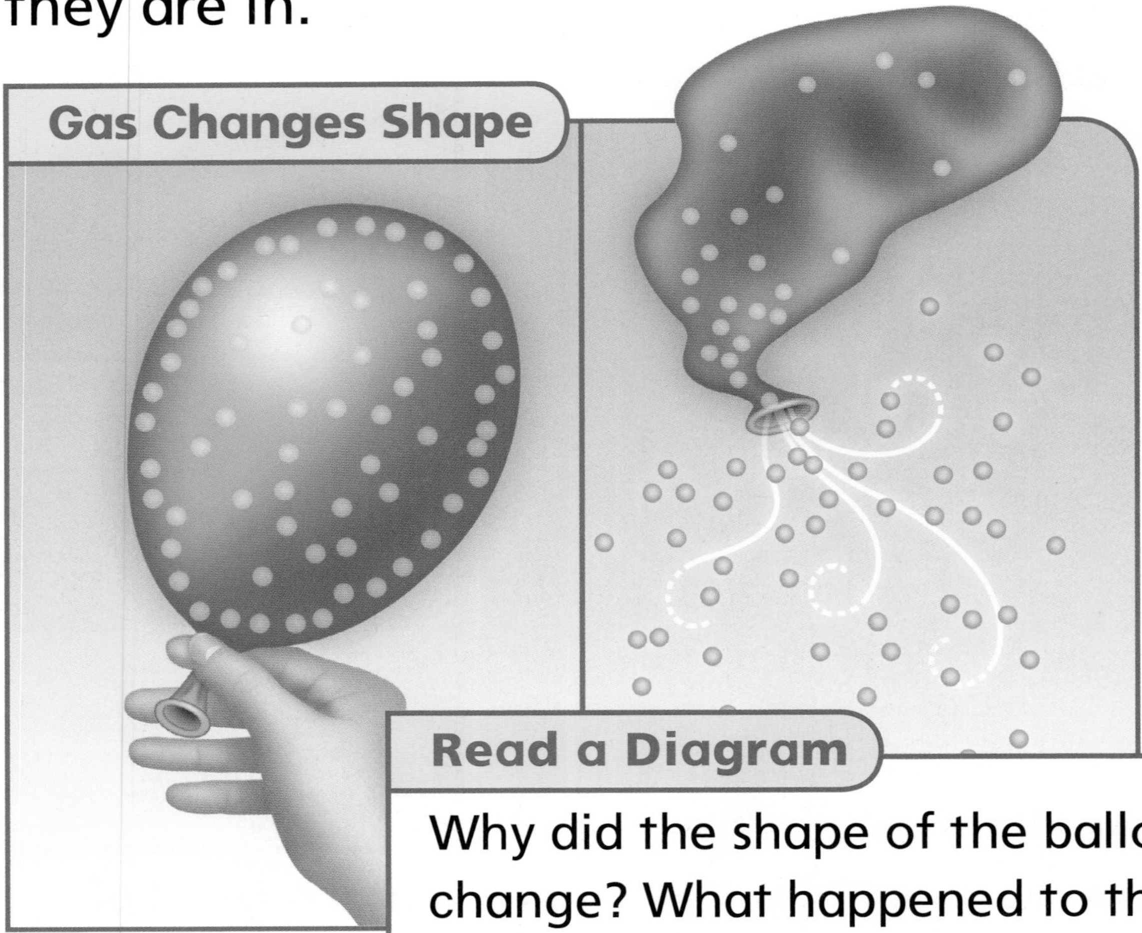

Read a Diagram

Why did the shape of the balloon change? What happened to the gas?

LOG ON *Science in Motion* Watch gas changing shape at www.macmillanmh.com

FACT Air is matter. It has mass and takes up space.

The air we breathe is made up of different gases.

You can not see these gases, but you can feel them. Air can feel hot or cold. It can also move.

Quick Lab

Observe how a tissue stays dry in water.

Even though you can not see the air, it helps this ribbon stay up. ▶

✔ How can you describe gas?

Think, Talk, and Write

1. **Predict.** What would happen to the gas in a balloon if it had a hole?
2. Write about what happens to a liquid when it spills on the floor.

Art Link

Draw a picture that includes solids, liquids, and gases.

LOG ON e-Review Summaries and quizzes online at www.macmillanmh.com

I Read to Review

My Matter Book

What is matter?
There are states it could be.
Solid, liquid, or gas,
what matter do you see?

A rock is a solid.
It can be rough or smooth.
It keeps its shape,
even if moved.

Water is a liquid.
It moves fast or slow.
It changes its shape
wherever it goes.

Air is a gas,
you can not see.
When let out,
it spreads evenly.

Vocabulary

Match each word to a picture.

balance
gas
liquid
solid

1
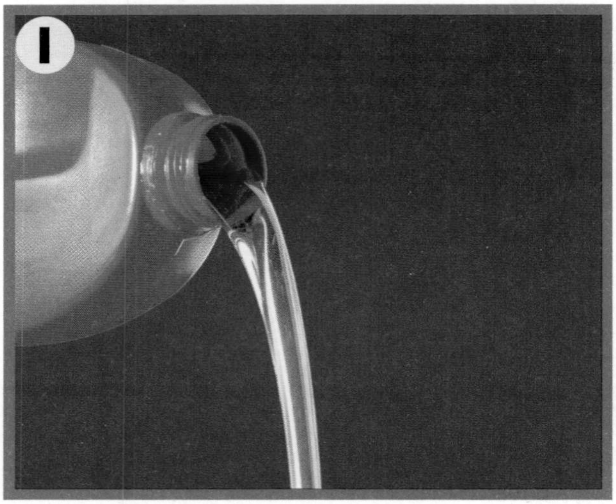

2

3
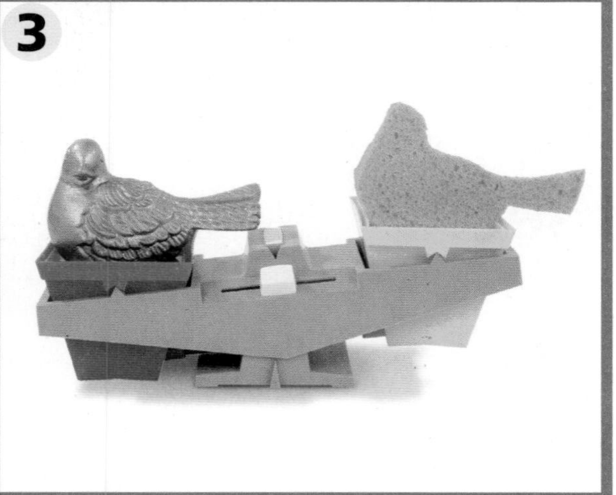

4

Science Skills and Ideas

Answer the questions below.

5. Describe the different textures you see on these puppets.

6. Measure. How can you measure mass?

7. Predict. If you blow into a balloon, what will happen?

8. Describe the properties of the liquids below.

The Big Idea

9. What are things made of?

CHAPTER 10

Changes in Matter

The Big Idea

How can matter change?

Key Vocabulary

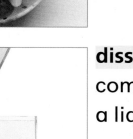

mixture two or more different things put together (page 334)

dissolve to completely mix into a liquid (page 336)

freeze to change from a liquid to a solid (page 342)

melt to change from a solid to a liquid (page 343)

More Vocabulary

burn, page 330

evaporate, page 344

Lesson 1

Matter Can Change

Look and Wonder

In Japan, people change paper into different shapes. It is called origami. How has this paper been changed?

Explore

Inquiry Activity

How can you change some solids?

What to Do

1. **Observe.** Describe a piece of paper, aluminum foil, and a tissue. How do they look and feel?
2. **Investigate.** How can you change each solid?
3. **Communicate.** What changed about each solid? What stayed the same? Make a chart of the changes.

Explore More

4. **Put Things in Order.** Put a few drops of water on the materials. Write what happens first, next, and last.

You need

paper

aluminum foil

tissues

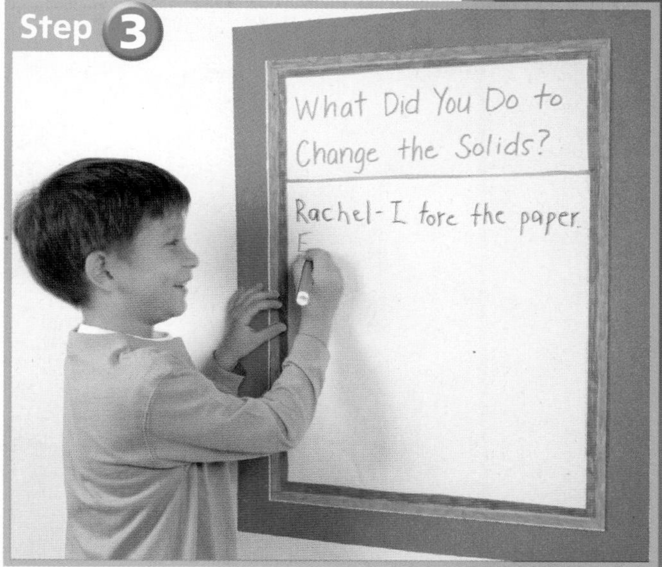

Step 3
What Did You Do to Change the Solids?
Rachel- I tore the paper.

Read Together and Learn

Vocabulary

burn

How can matter change?

You can bend, fold, or tear some solids. The solid looks different. But it is still made out of the same thing.

Sometimes matter can change into something else. When you **burn** matter, heat and air change its properties.

Shaping Clay

Read a Photo

How has this clay been changed?

When you cut paper, the pieces are smaller. But it is still paper.

When paper burns, it turns into ash. It is no longer paper.

Quick Lab

Investigate how sunlight can change paper.

How can you change an apple?

Think, Talk, and Write

1. **Problem and Solution.** Your homework got crumpled. How can you change it back to how it was?
2. Write about how burning changes the properties of paper.

Art Link

Try to make a boat using paper.

LOG ON e-Review Summaries and quizzes online at www.macmillanmh.com

Lesson 2

Making Mixtures

Look and Wonder

Some mixtures are made of solids. What do you see in this mixture?

Explore

Inquiry Activity

Can you take a mixture apart?

What to Do

1. **Observe.** Mix sand and birdseed together. What do you notice?
2. **Predict.** How could you separate your mixture?
3. **Investigate.** Pour your mixture into a strainer. What happens?
4. **Put Things in Order.** Write what happens first, next, and last.

Explore More

5. **Investigate.** Could you separate the mixture if you added water? Why or why not? Try it!

You need

cup

sand

birdseed

strainer

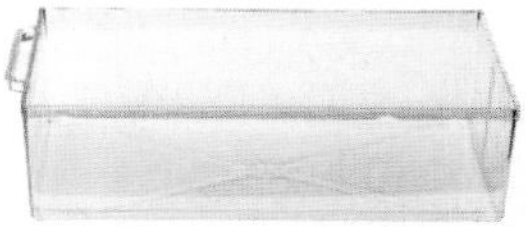
clear bin

Step 3

Read Together and Learn

Vocabulary

mixture

dissolve

SCIENCE QUEST Explore making mixtures with the Junior Rangers.

What is a mixture?

A **mixture** is two or more different things put together.

When you mix some solids, they do not change. You can see the parts of the mixture. You can take them apart.

It is easy to take this mixture of solids apart. ▶

When you mix solids with water, some objects will float. Others will sink.

Sometimes you can pick the solids out of the water.

✔ **What are some ways to separate a solid mixture?**

Quick Lab

Investigate what sinks and what floats in water.

Read a Photo

Why do you think the red balls float and the marbles sink?

Sink and Float

What are some other mixtures?

Some mixtures are hard to take apart. When you mix a solid and a liquid, some solids **dissolve**, or mix completely into the liquid. Some liquids mix completely, too.

▼ **Drink mix will dissolve in water.**

▼ **You can not take food coloring out of water.**

FACT A solid that dissolves in water may not dissolve in other liquids.

Some liquids do not mix together. Oil and water do not mix. The two liquids stay apart.

Oil floats on top of water. ▶

✔ What mixtures are not easy to take apart?

Think, Talk, and Write

1. **Main Idea and Details.** Describe a mixture made of two solids.
2. Write and draw about a solid that you can mix with water.

Music Link

Listen to "Mix and Change" on the Science Songs CD. Add your own verse.

LOG ON e-Review Summaries and quizzes online at www.macmillanmh.com

Mix It Up

Look at the picture below. What do you think happened? What are some parts of this mixture?

Write a Story

Write a story about this mixture. Can you take it apart with your hands or do you need to use a tool?

Remember

Use words to tell how something looks.

LOG ON e-Journal Write about it online at www.macmillanmh.com

Trail Mix Recipe

Carrie made trail mix. She used this recipe. She mixed everything together.

Carrie's Trail Mix

2 cups dried fruit

1 cup nuts

1 cup raisins

Write a Number Sentence

Make your own trail mix. Write a number sentence to show how many cups of each food you used in your mix.

Remember

A number sentence helps you solve a problem.

Lesson 3

Heat Can Change Matter

Look and Wonder

In winter, frozen water falls from the sky as snow. What can happen to the snow on a sunny day?

Explore

Inquiry Activity

How can heat change ice?

What to Do

1. **Measure.** Take the temperature of a cup of cold water. Then take it of a cup of warm water.
2. **Predict.** In which cup will an ice cube melt faster? Add one ice cube to each cup.
3. **Put Things in Order.** Which ice cube melted first? Why?
4. **Measure.** Take the temperature of the water again. Did it change?

Explore More

5. **Investigate.** Try this activity again. Did you get the same results?

You need

cups of water

thermometer

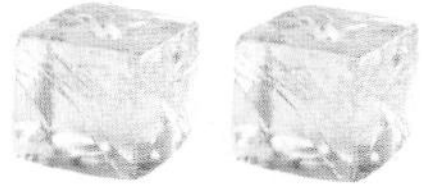
ice

Step 2

Read Together and Learn

Vocabulary

freeze

melt

evaporate

SCIENCE QUEST Explore how heat can change matter with the Junior Rangers.

How can solids and liquids change?

You can change liquid water to a solid. When a liquid gets very cold, it can freeze.

To **freeze** means to change from a liquid to a solid.

▼ **When you put liquid water in a freezer, it turns into a solid.**

Forms of Water

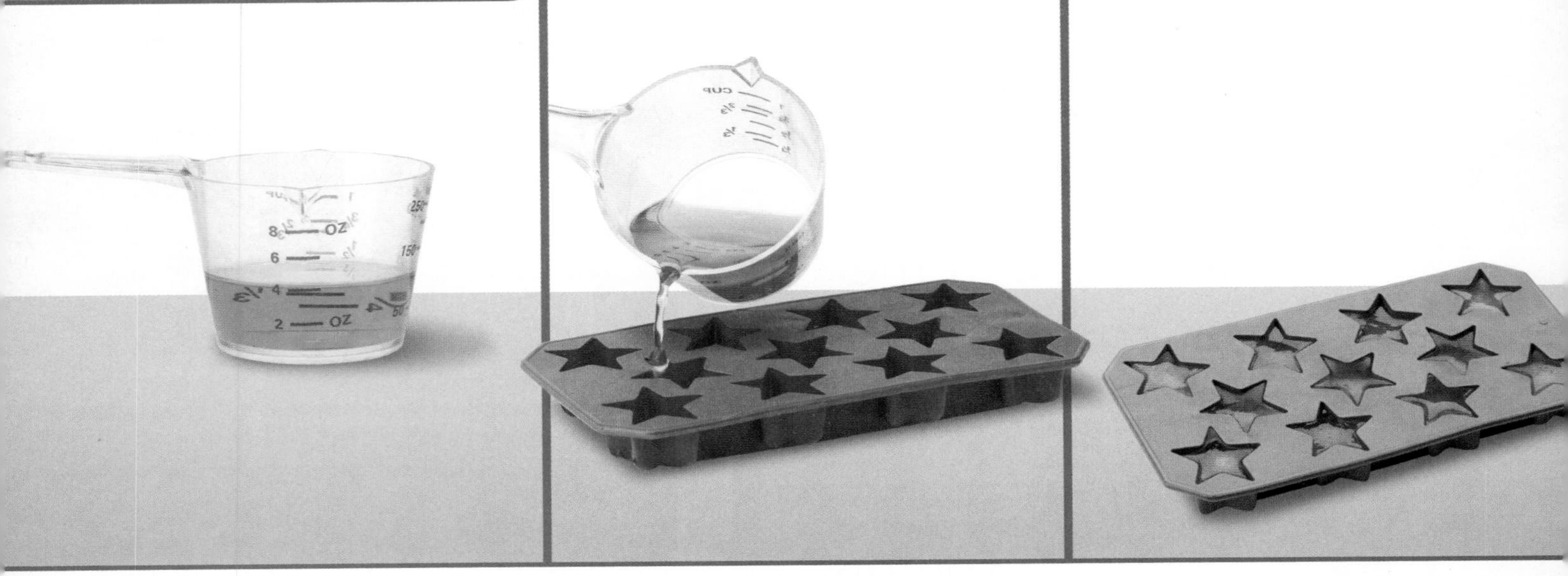

FACT Freezing is taking away heat, not adding cold.

You can change frozen water back into a liquid. When heat is added to a solid, it melts.

To **melt** means to change from a solid to a liquid. Different solids melt with different amounts of heat.

Quick Lab

Use a balance to see if water has the same mass as ice.

▼ **When the ice melts, the water in the cup is the same amount of water that you started with.**

Read a Photo

How did the water change? How did it stay the same?

LOG ON *Science in Motion* Watch water change states at www.macmillanmh.com

✓ **What happens when you add heat to a solid?**

How can liquids and gases change?

When water gets warm, some water evaporates. To **evaporate** is to go into the air as a gas.

The more heat you add to water, the faster it will evaporate. Water vapor is water as a gas.

▲ **At one time, this pond was full of water.**

▲ **Some water evaporated and the pond got smaller.**

When water vapor gets cool, it changes into a liquid.

When water vapor in the air touches a cool glass, it turns back into water. That is why your glass is sometimes wet on the outside.

✓ **What happens when water is heated?**

▲ **You can see water drops on this pitcher.**

Think, Talk, and Write

1. **Summarize.** How can water change from a solid to a liquid?
2. Write about how water can change to a gas.

Health Link

Make frozen juice pops from fruit juice.

LOG ON e-Review Summaries and quizzes online at www.macmillanmh.com

Real World

Reading in Science

Hot Stuff

Do you want your hot chocolate to stay hot? You need to choose the right kind of cup.

Heat moves from a hotter place to a cooler place. Some heat leaves through the top of your cup. Heat also leaves through the sides of your cup.

Heat travels quickly through the sides of a ceramic mug. Heat travels slowly through the sides of a foam cup.

A foam cup holds heat better than a ceramic mug. Air in the foam keeps the heat from getting out.

Talk About It

Problem and Solution. What kind of cup helps your drink stay warm? Why?

AMERICAN MUSEUM of NATURAL HISTORY

I Read to Review

My Mixtures Book

You can make
many mixtures.

Some are easy to take
apart. Others are not.

Fruit salad can
be taken apart.
Smoothies can not.

Apples can be
taken out of a pie.
Pumpkin can not.

CHAPTER 10 Review

Vocabulary

Use each word once to complete the sentences.

dissolve
freeze
melt
mixture

1. Some solids completely mix, or ______ in water.

2. Two or more things put together make a ______.

3. To change from a solid to a liquid means to ______ something.

4. To change from a liquid to a solid means to ______ something.

Science Skills and Ideas

Answer the questions below.

5. What can happen if you mix a solid with a liquid?

6. **Problem and Solution.** How can you stop a frozen juice pop from melting?

7. **Put Things in Order.** Tell how burning changes paper.

8. How many different ways can you change paper?

The Big Idea

9. How can matter change?

Baker

Do you love to make cookies? You could become a baker. A baker makes breads, cookies, cakes, and other foods to sell. Some bakers have their own bakery. Others work for big companies.

You must understand the science of baking to become a baker. Bakers learn that when different foods are mixed, they can change.

baker

More Careers to Think About

chemist

cement truck operator

UNIT F

Motion and Energy

We can not see the wind, but we can see what it moves.

Literature
Poem

FOR A QUICK EXIT

by Norma Farber

For going up or coming down,
in big department stores in town,
you take an escalator.
(They come in pairs.)
Or else an elevator.
(Also stairs.)

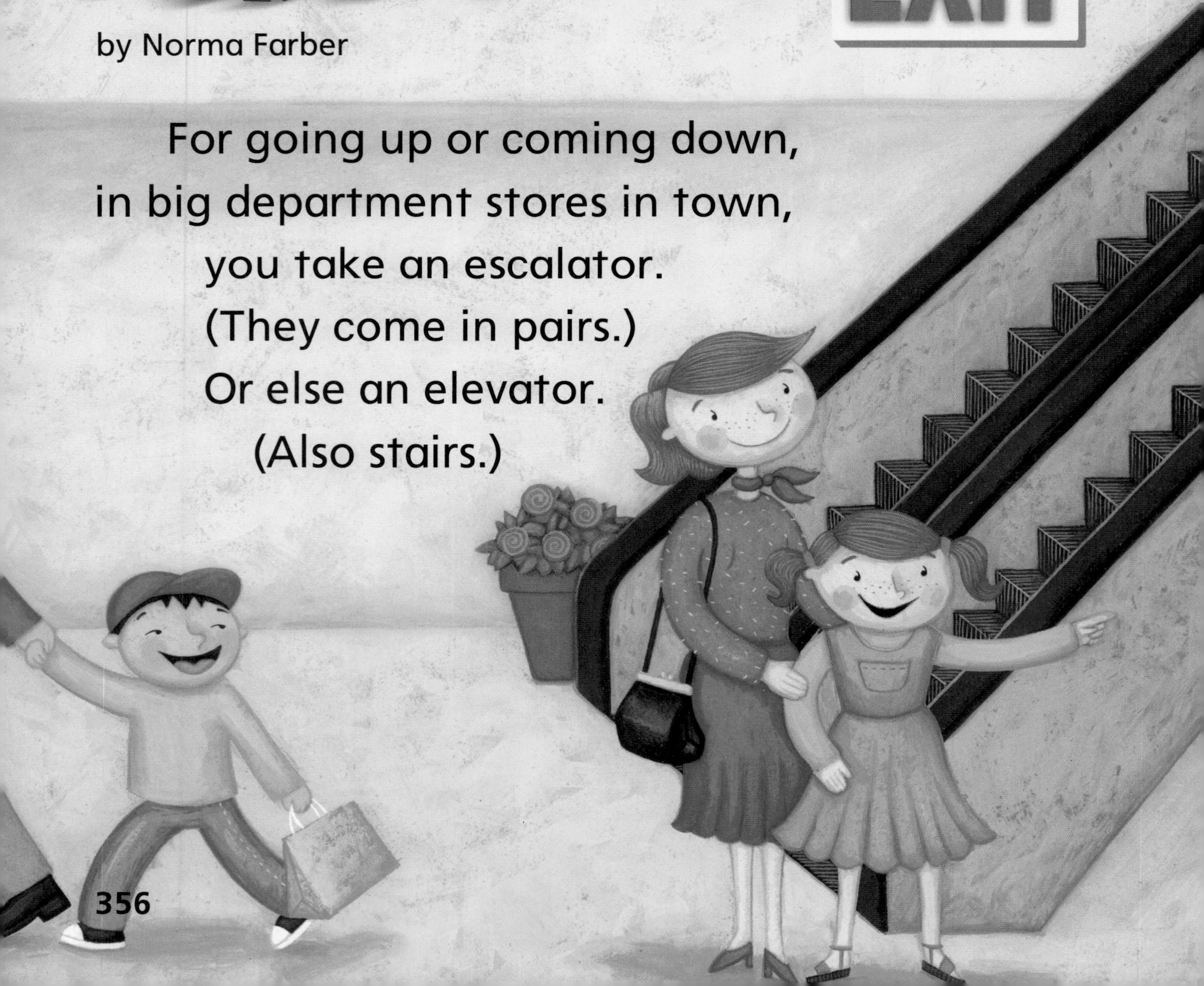

I wish storekeepers would provide a SLIDE!

Talk About It

What are some ways to get from one floor to another in a building?

CHAPTER 11

On the Move

The Big Idea

How can you make things move?

Key Vocabulary

push a force that moves something away from you (page 368)

pull a force that moves something closer to you (page 368)

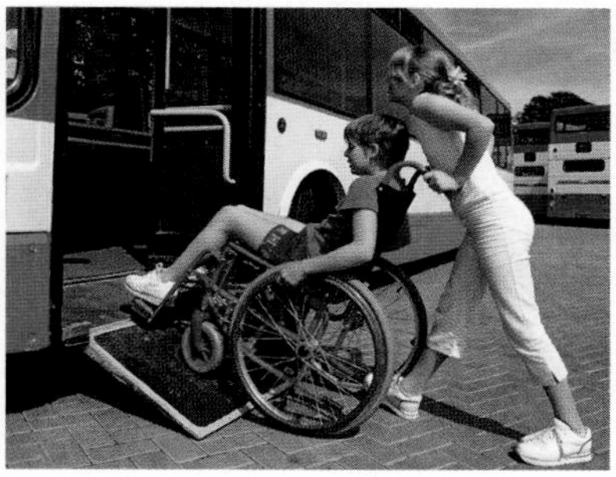

ramp a slanted surface that you can use to move things up or down (page 377)

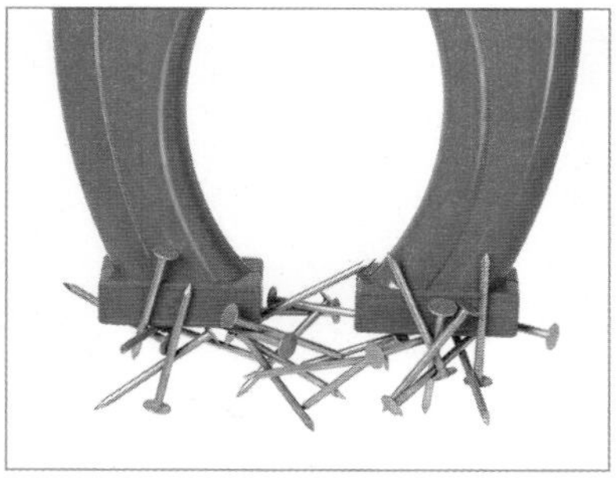

magnet something that can pull, or attract, some objects with metal in them (page 382)

More Vocabulary

position, page 362

motion, page 364

speed, page 365

force, page 368

gravity, page 369

friction, page 371

simple machine, page 374

pulley, page 375

lever, page 376

poles, page 384

repel, page 385

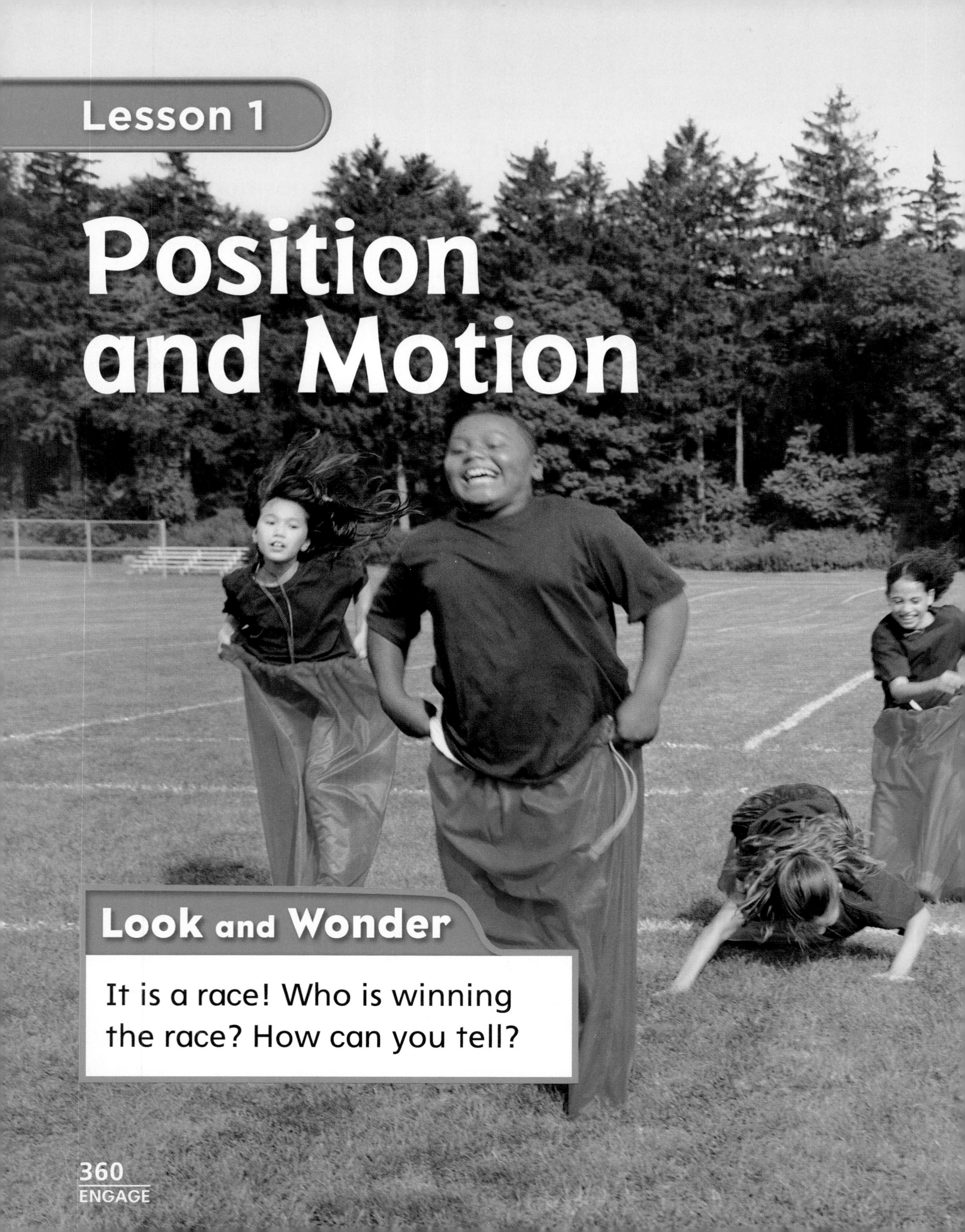

Lesson 1

Position and Motion

Look and Wonder

It is a race! Who is winning the race? How can you tell?

Explore

Inquiry Activity

How do you know something moved?

What to Do

1. Put three objects on a table.
2. **Observe.** Look closely at the objects. Where are they on the table?
3. Cover your eyes. Have your partner move one object.
4. **Infer.** Open your eyes. Which object did your partner move? How can you tell?

Explore More

5. **Investigate.** Can making a map of the table and objects help you find out which object moved? Try it.

You need

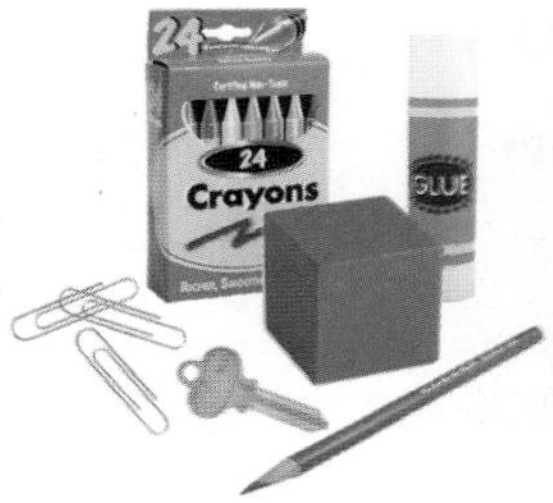

classroom objects

Read Together and Learn

Vocabulary

position
motion
speed

How can you tell where something is?

Have you ever told a friend where something is? You probably described the object's position.

Position is the place where something is located.

Find Things at a Fair

Position tells you if one thing is close to or far away from another thing. Position can tell you if an object is over, under, right, or left.

What other words can you use to describe an object's position?

Read a Photo

Where are things located at this fair? Use position words.

How do things move?

Objects can move in many ways. **Motion** is a change in an object's position.

Things can move forward, backward, or in a circle. They can even zigzag!

Quick Lab

Make a ball move in a curvy line and in a straight line.

▲ This car drives down a curvy road.

▲ This airplane moves in a straight line.

Speed is how fast or slow something moves. Different objects move at different speeds.

A rocket ship moves much faster than an airplane. ▶

✔ How can you tell if one object is moving faster than another?

Think, Talk, and Write

1. **Compare and Contrast.** How are a rocket ship and an airplane alike? How are they different?
2. Write about the different ways you can move a ball.

Health Link

Have a classmate use position words. Move to the location he or she describes.

LOG ON e-Review Summaries and quizzes online at www.macmillanmh.com

Lesson 2

Pushes and Pulls

Look and Wonder

This girl is climbing a rope. How does she move up?

Explore

Inquiry Activity

How can you make something move?

What to Do

1. Fold an index card.
2. **Investigate.** Try different ways to make the card move. How can it move?
3. **Observe.** What changes about the card? What stays the same?

Explore More

4. **Infer.** Do you think a tissue will move in the same way as the card? Why or why not? Try it.

You need

index card

tissue

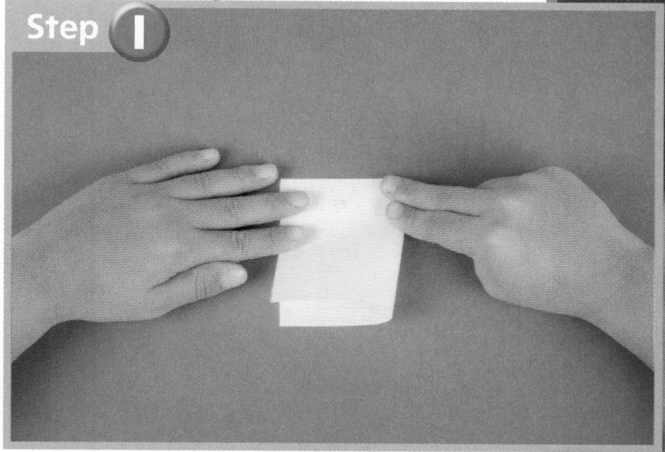

Read Together and Learn

Vocabulary

force

push

pull

gravity

friction

What makes things move?

Things can not move on their own. You have to use force to move them. A **force** is a push or a pull that makes an object move.

A **push** moves the object away from you. A **pull** moves it toward you.

◀ **This girl pushes the basketball away from her.**

▲ **This boy pulls the bag of basketballs toward him.**

Gravity is the force that pulls things toward Earth.

When you jump up, gravity pulls you back down. If you let go of something, gravity pulls it to the ground.

What things do you push and pull every day?

Jumping Rope

Read a Photo

Will this girl stay up in the air? Why or why not?

LOG ON *Science in Motion* Watch how gravity works at www.macmillanmh.com

Gravity pulls this egg to the ground. ▼

How are forces different?

The size of a push or pull moves things differently. A small push can move a light object. A bigger push can move a heavy object.

A big push also makes an object move faster and farther than a small push.

Quick Lab

Investigate how much force you need to slide a checker piece.

▲ **This boy uses a big force to push the golf ball far away.**

▲ **This boy uses a small force to push the golf ball a short way.**

Friction is a force that slows things down. **Friction** is two things rubbing together.

Have you ever dragged your feet to slow down on a swing? That is friction.

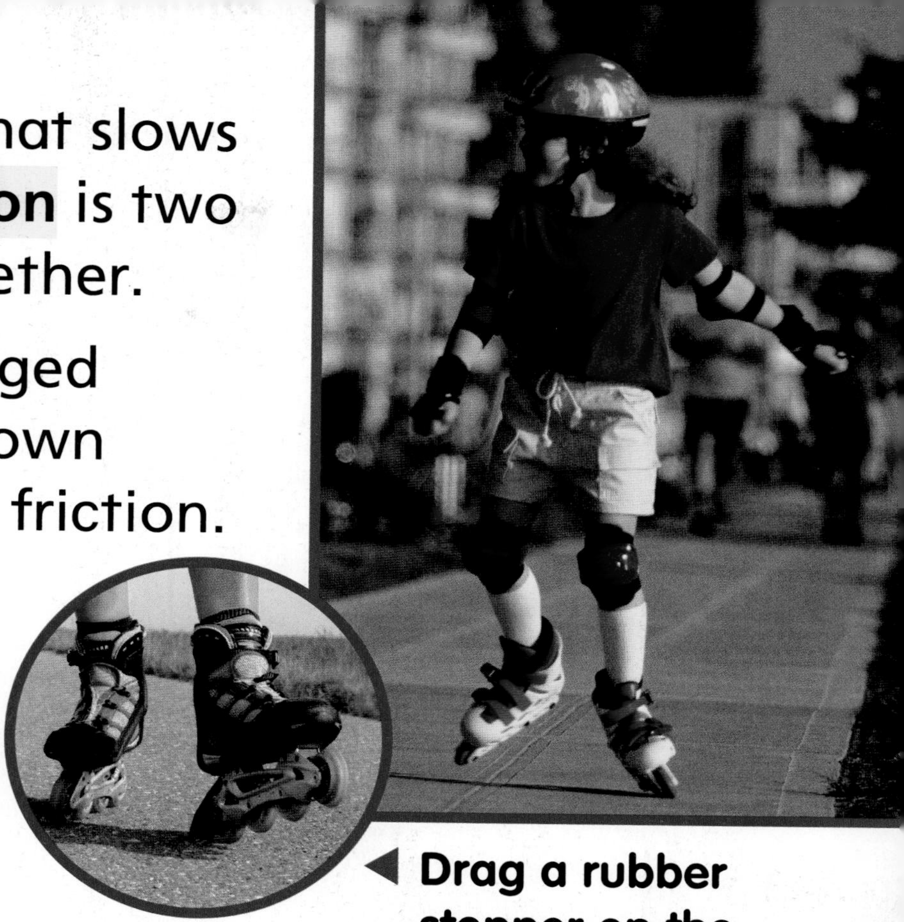

Drag a rubber stopper on the ground. Friction makes you stop.

What could make something move slower?

Think, Talk, and Write

1. Cause and Effect. What makes things fall to the ground?
2. Write about what can happen if you use a big push on a light object.

Social Studies Link

Describe a game that people play with a ball. What forces make the ball move?

LOG ON e-Review Summaries and quizzes online at www.macmillanmh.com

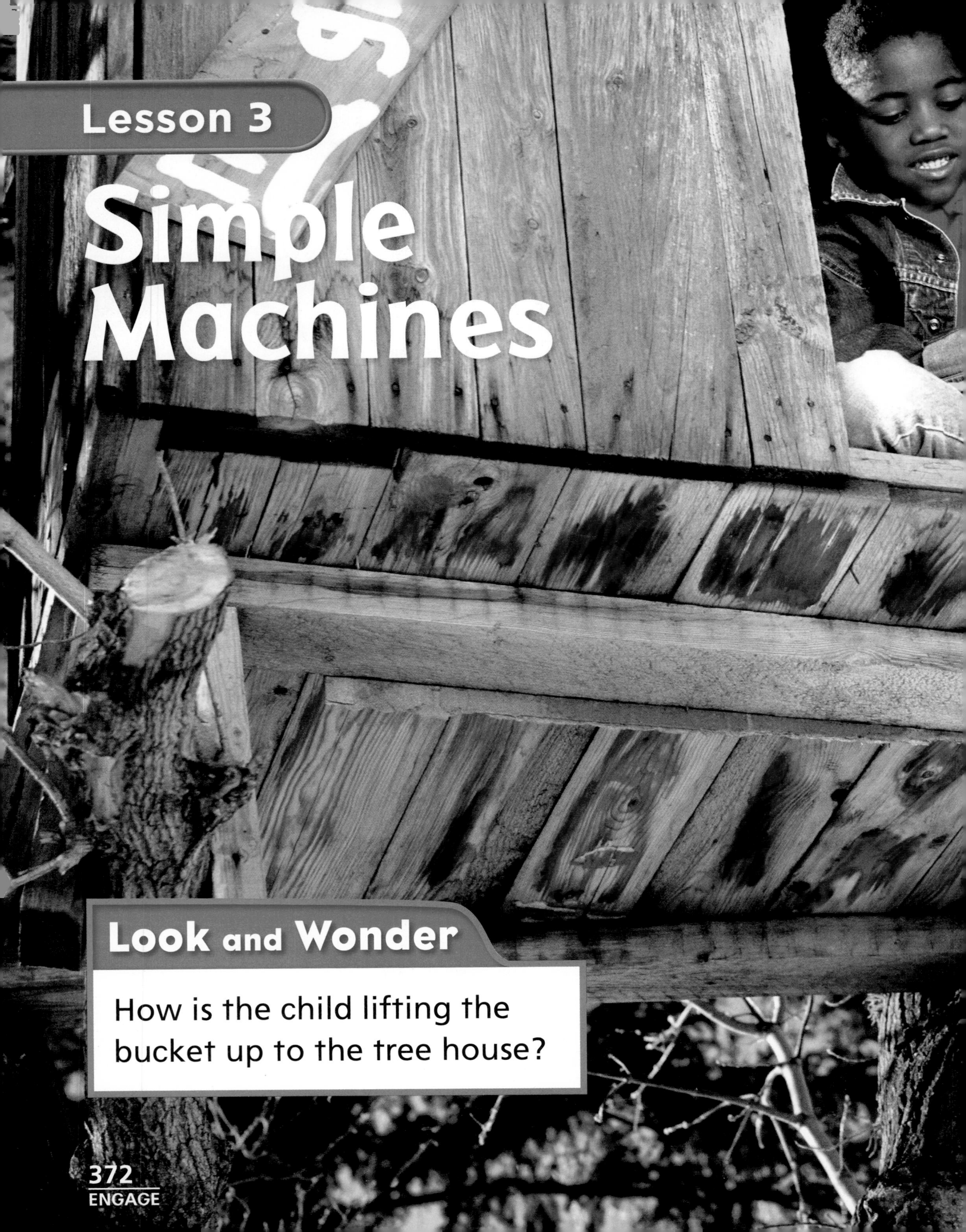

Lesson 3

Simple Machines

Look and Wonder

How is the child lifting the bucket up to the tree house?

Explore

Inquiry Activity

How can you lift a book?

What to Do

1. Tie a string around a book. Do not cut the end of the string.
2. Put the book on the floor. Pull the string through a drawer handle.
3. **Predict.** What will happen if you pull down on the string?
4. **Investigate.** Pull down on the string. What happens to the book?

Explore More

5. **Infer.** When would it be useful to lift an object this way?

You need

string

book

Step 2

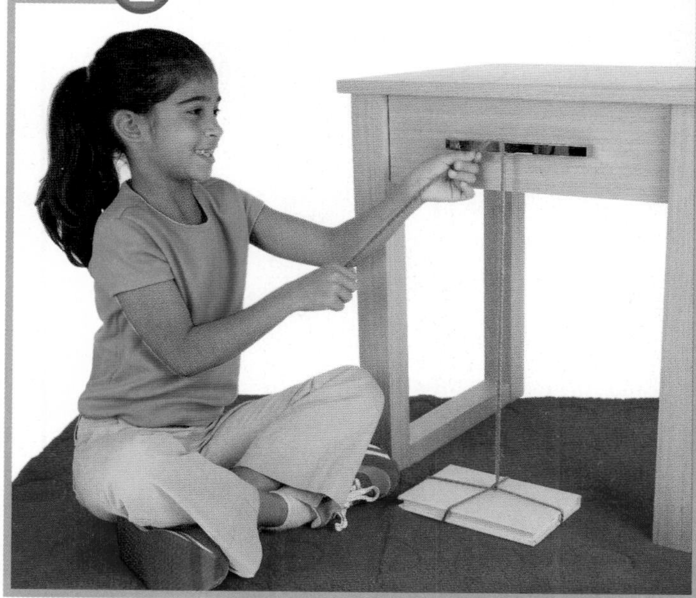

Read Together and Learn

Vocabulary

simple machine
pulley
lever
ramp

What are simple machines?

Sometimes things are hard to move. You might need help moving them.

A **simple machine** is a tool that can make it easier to move things. Simple machines help people do work.

Using Simple Machines in the Yard

Read a Photo

What are these rakes helping the children do?

One kind of simple machine is a pulley. A **pulley** is a rope that moves over a wheel.

A pulley makes it easier to lift heavy objects. It also helps move things to high places.

✓ **How can we use pulleys?**

The parts of a pulley work together to help us get flags up to the top of a flagpole.

FACT Pulleys can be up high or down low.

What are levers and ramps?

A lever is another kind of simple machine.

A **lever** is a bar that balances on a point. It moves like a seesaw. A lever makes it easier to move something.

Quick Lab

Use a ruler and a marker. Make a lever lift a book.

lever

▼ **An oar is a lever that helps you move a boat.**

A ramp is also a simple machine. A **ramp** is a slanted surface. You can use it to move things up or down.

People sometimes use ramps to move things that are hard to get up stairs.

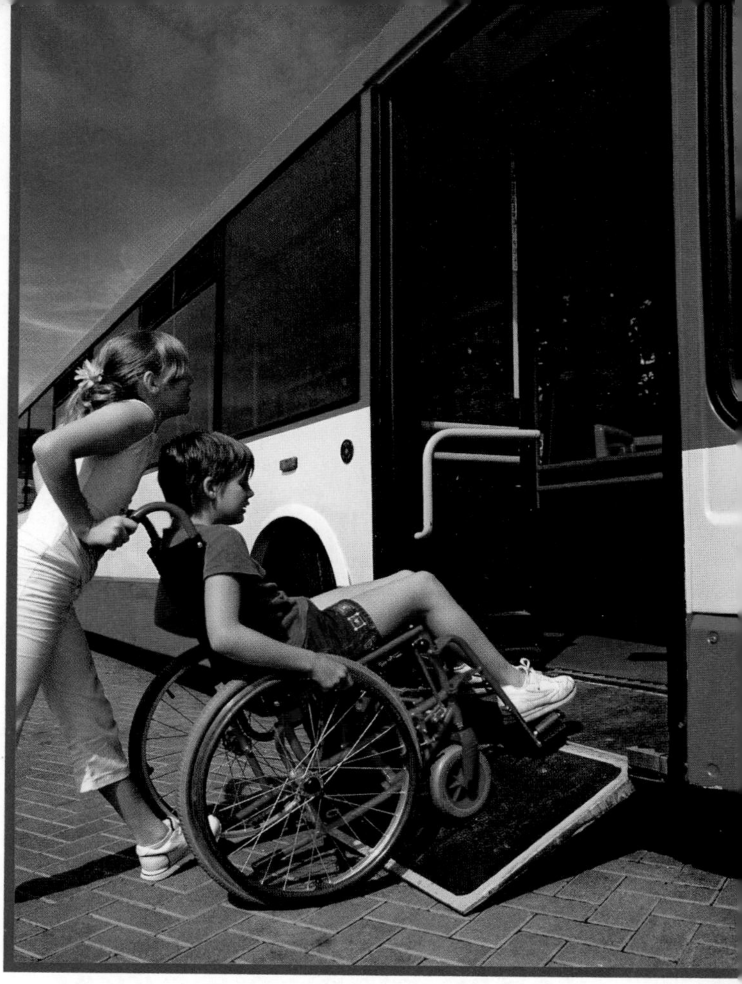

A ramp helps people in wheelchairs get onto buses.

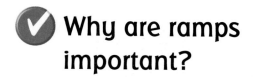

Why are ramps important?

Think, Talk, and Write

1. **Classify.** What kind of simple machine is a seesaw?
2. Write about how the different parts of a pulley help it work.

Art Link

Draw a ramp and how it can help you.

LOG ON e-Review Summaries and quizzes online at www.macmillanmh.com

MOVING UP

Pulleys and ramps are useful simple machines. They can help you get up to a high place. They can also help you get down from one.

Pulleys and ramps are used in many everyday things. Most elevators use a pulley on a motor. The pulley lifts an elevator up and down.

Escalators are moving ramps with motors. The motor uses a belt to pull steps around a ramp.

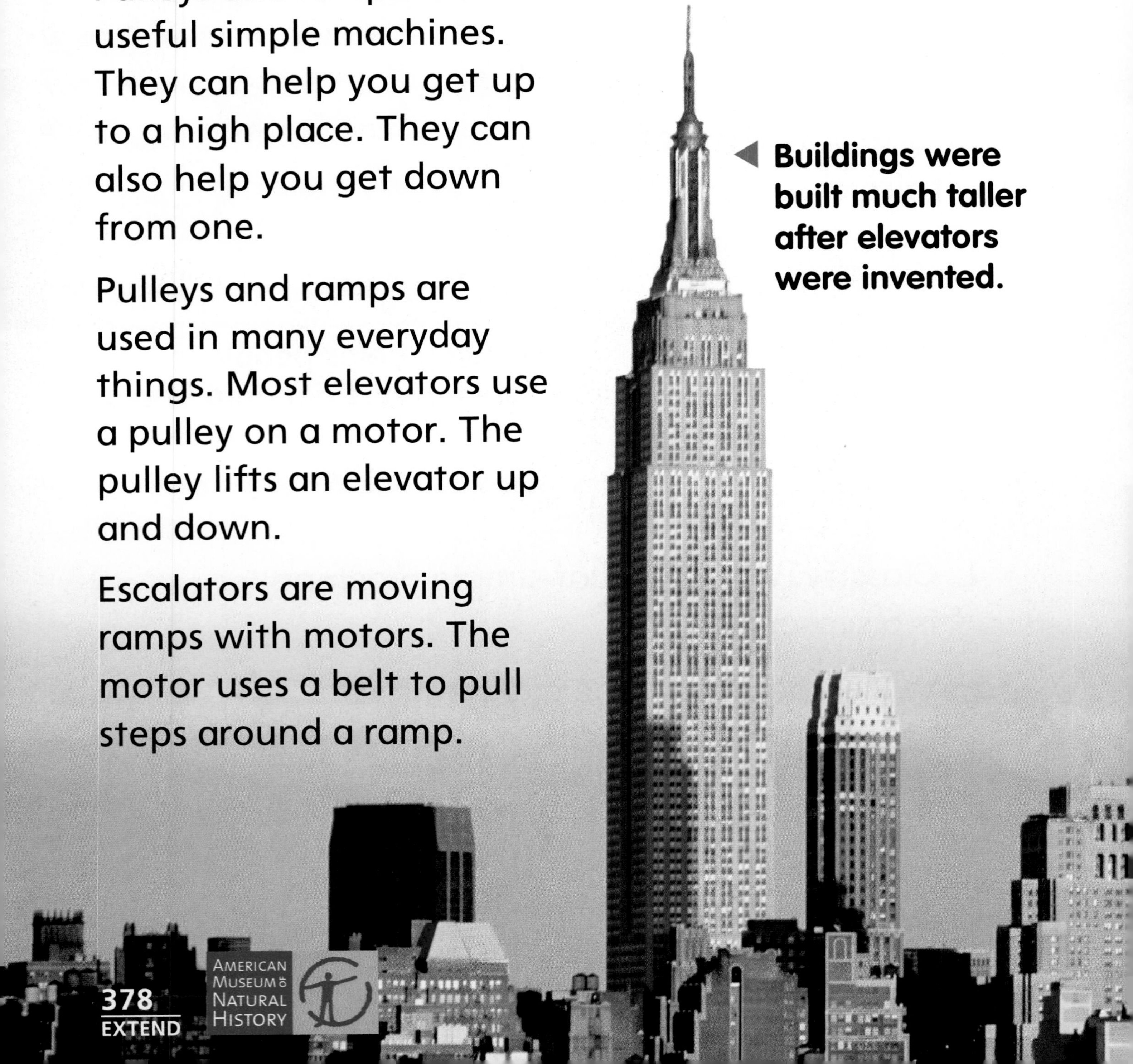

Buildings were built much taller after elevators were invented.

AMERICAN MUSEUM OF NATURAL HISTORY

The first public elevator was built in 1857.

The first modern escalator was built in 1921.

Talk About It

Classify. Make a list of pulleys and ramps you have used. How did you use them?

Lesson 4

Magnets

Look and Wonder

Magnets pull things toward them. Where are the magnets on this train?

Explore

Inquiry Activity

What will a magnet pull?

What to Do

1. **Predict.** Put objects that you think a magnet will pull in one pile. Put objects it will not pull in another pile.
2. **Investigate.** Put the magnet close to different objects. What happens?
3. **Classify.** Which objects were pulled by the magnet? Which objects were not?

Explore More

4. **Infer.** What kinds of objects do magnets pull?

You need

magnet

classroom objects

Step 2

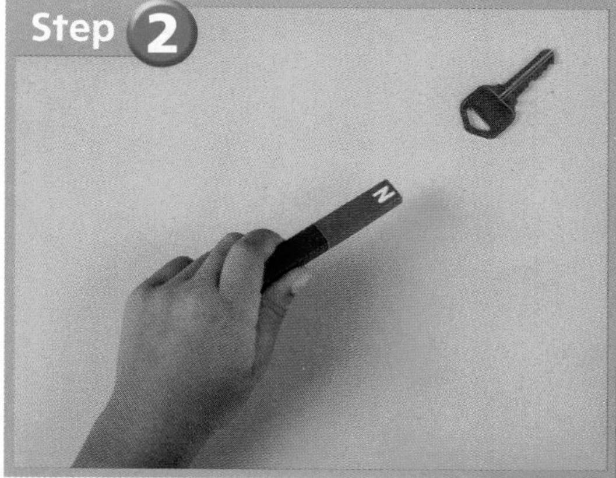

Read Together and Learn

Vocabulary

magnet
poles
repel

What is a magnet?

Some things stick together with tape or glue. A magnet does not need those things to stick to something.

A **magnet** pulls, or attracts, some kinds of objects.

Which objects on the table will this magnet attract? ▼

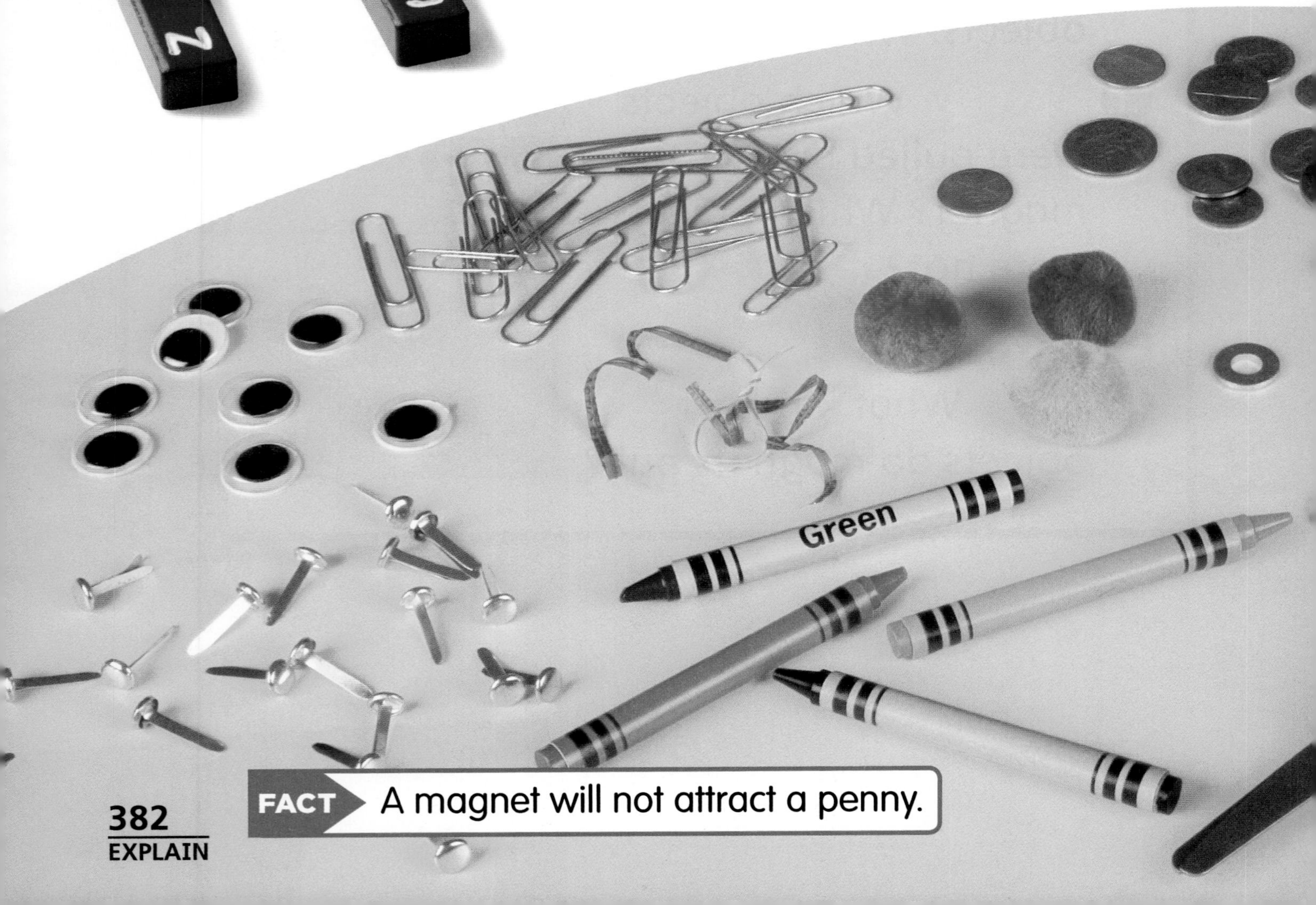

FACT A magnet will not attract a penny.

Magnets attract things that have iron in them. Iron is a kind of metal.

Magnets do not attract things that are made of plastic, rubber, or cloth.

✔ **What kind of objects will a magnet attract?**

Magnet Chart

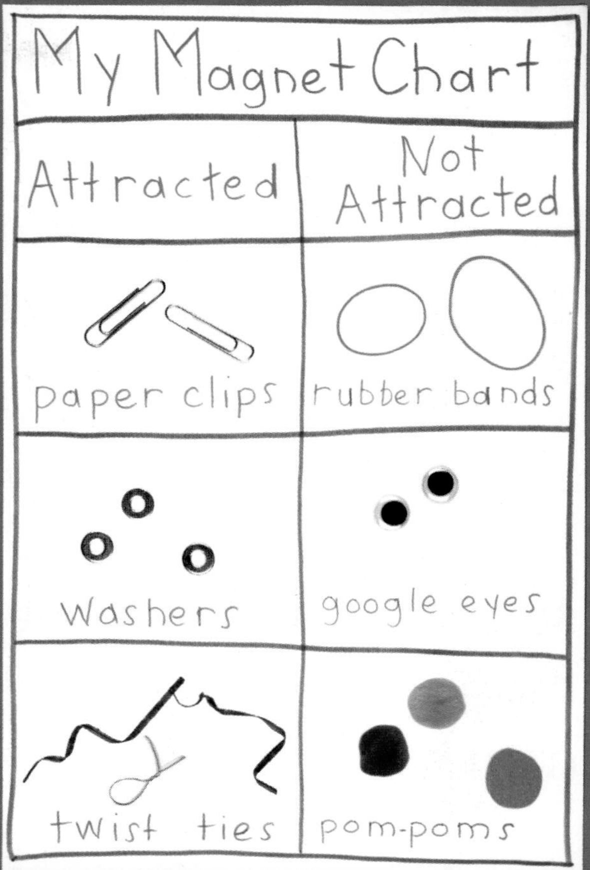

My Magnet Chart
Attracted
Not Attracted
paper clips
rubber bands
washers
google eyes
twist ties
pom-poms

Read a Chart

Did the magnet attract rubber bands? Why or why not?

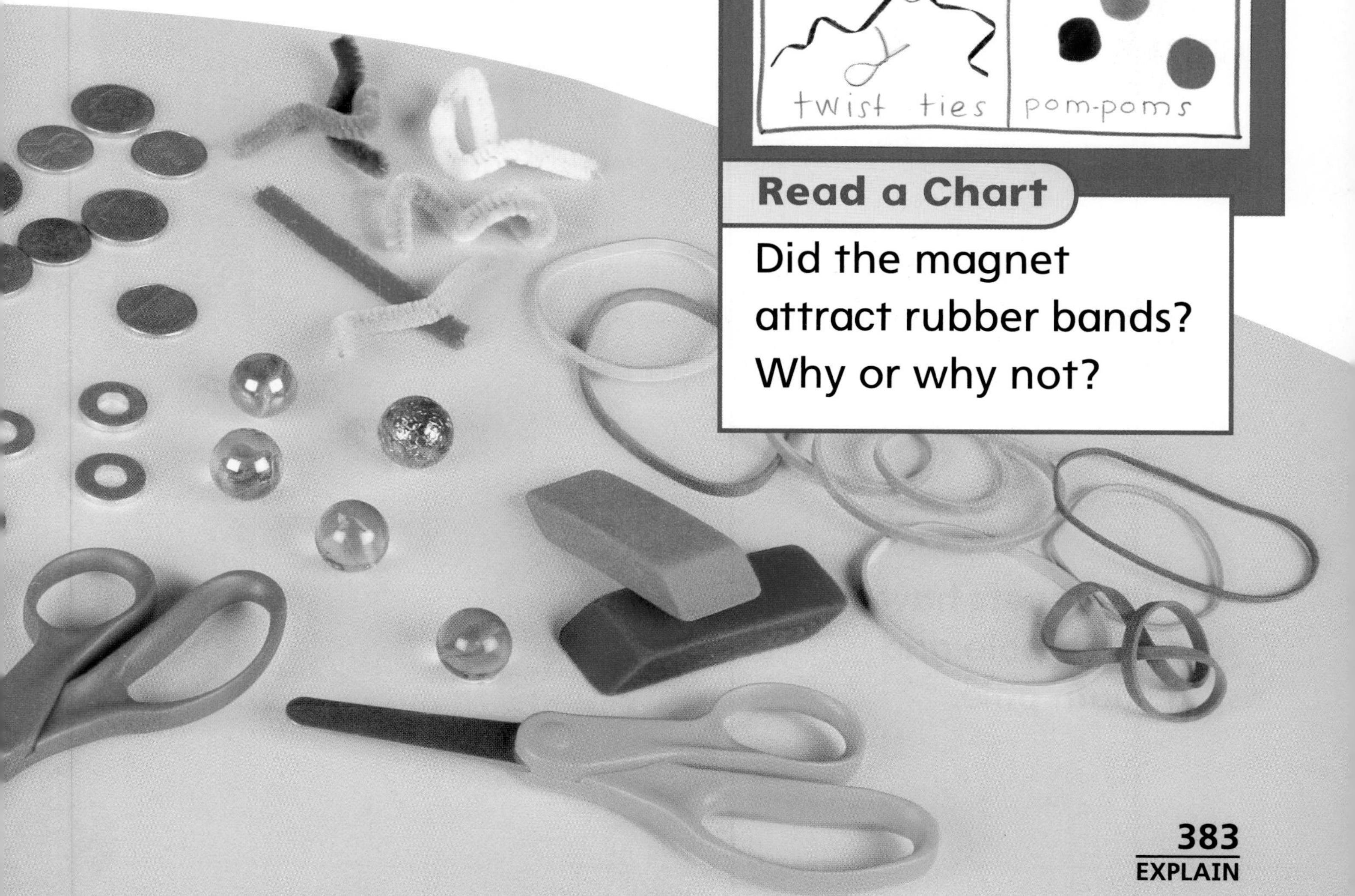

What are a magnet's poles?

Every magnet has two poles. **Poles** are where the magnet's pull is strongest. The **N** shows the north pole. The **S** shows the south pole.

If you hold the north pole of one magnet up to the south pole of another magnet, the poles will attract.

Quick Lab

See if a magnet can pull through paper, water, or your hand.

Magnets can be different shapes.

All magnets have a north pole and a south pole.

If you put two north poles or two south poles next to each other, they will repel one another. **Repel** means to push away.

These magnets have red north poles and blue south poles. ▶

✔ Why is there space between some magnets on this pencil?

Think, Talk, and Write

1. **Predict.** Can magnets pull through air, paper, water, and your hand?
2. Write or draw what happens when two like poles of a magnet are put together.

Social Studies Link

Find out some ways that people use magnets in your community.

LOG ON e-Review Summaries and quizzes online at www.macmillanmh.com

Fun with Magnets

Look at the picture below. Each car has a magnet at the end of it.

Write About It

Tell how the girl in this picture can use the magnets. Write a story about how you use magnets.

Remember
Tell what happens first, next, and last.

LOG ON e-Journal Write about it online at www.macmillanmh.com

Comparing Magnets

Molly had two magnets. She wondered which one would pick up more paper clips. She compared the amounts.

$3 < 12$

3 is less than 12

Compare

Use two different magnets. See which one picks up more paper clips. Compare the amounts.

> **Remember**
> The < symbol always points to the smaller number.

My Motion Book

Many things can move
with a little force.

A push or a pull
can change your course.

Magnets can move
some metal things around.

Gravity pulls things
to the ground.

CHAPTER 11 Review

Vocabulary

Use each word once to complete the sentences.

lever
magnet
motion
position
repel

1. An object's movement from one place to another is called _____.

2. An oar is a type of _____.

3. When an object is moving, its _____ changes.

4. When the two north poles of magnets face each other, they _____.

5. Objects made of iron will be attracted to a _____.

Science Skills and Ideas

Answer the questions below.

6. Use position words to describe where the clowns are in the picture below.

7. **Infer.** What will happen if this girl drags her rollerblades' rubber stopper on the ground? Why?

8. **Classify.** What can ramps and levers help you do that you could not do without them?

The Big Idea

9. How can you make things move?

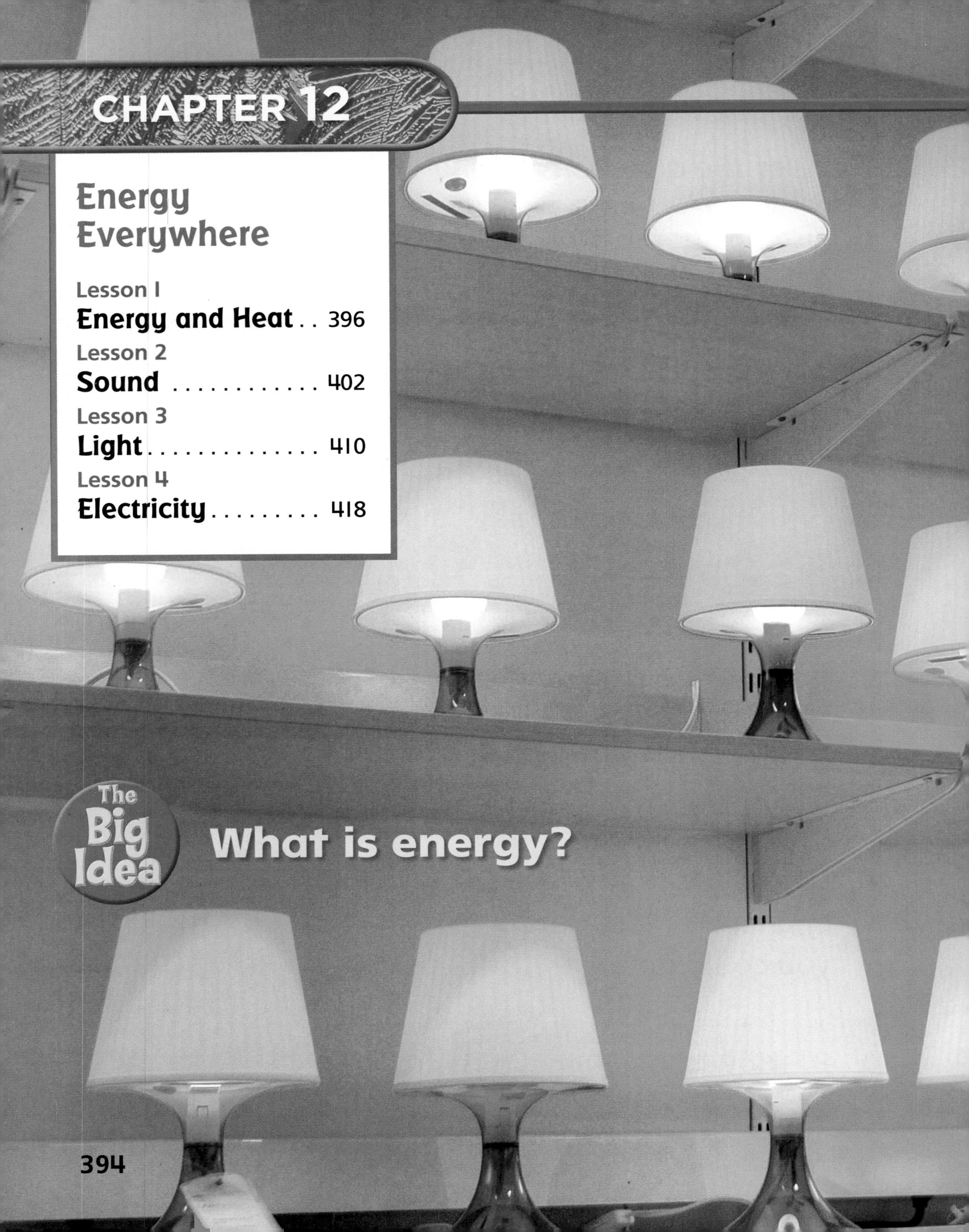

CHAPTER 12

Energy Everywhere

The Big Idea

What is energy?

Key Vocabulary

heat a form of energy that makes things warm (page 400)

vibrate to move back and forth quickly (page 404)

light a form of energy that lets you see (page 412)

electricity a form of energy that gives some things the power to work (page 420)

More Vocabulary

energy, page 398

pitch, page 407

Lesson 1

Energy and Heat

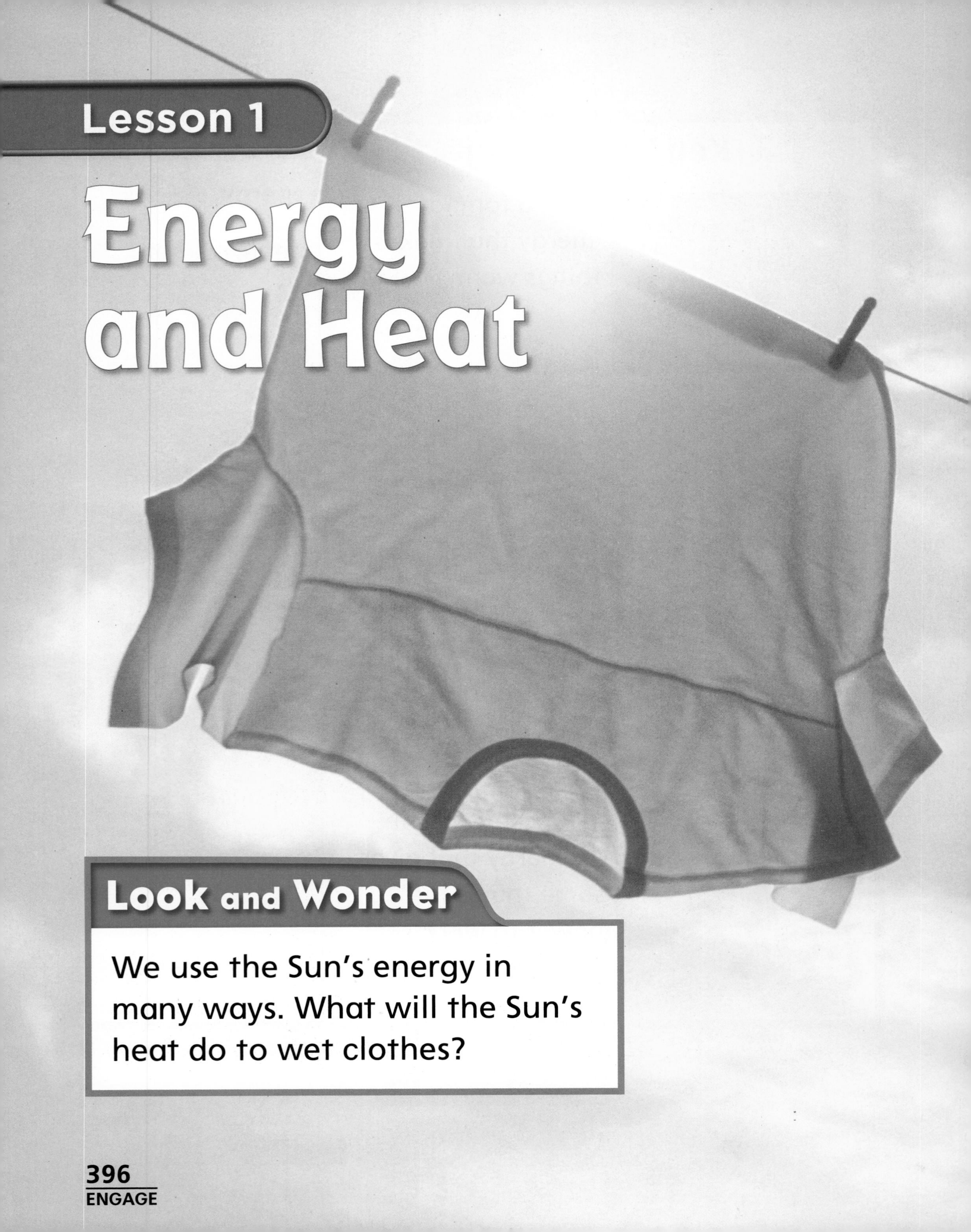

Look and Wonder

We use the Sun's energy in many ways. What will the Sun's heat do to wet clothes?

Explore

Inquiry Activity

How can heat change things?

What to Do

1. Put butter, an ice cube, and crayons on two plates. Place one plate in a warm place and the other in a cool place.
2. **Predict.** What will happen to the objects on each plate?
3. **Observe.** Wait ten minutes. How have the objects changed?
4. **Draw Conclusions.** What made the objects change?

Explore More

5. **Investigate.** Wait two hours. Check the objects. Have they changed?

You need

plastic plates

butter

ice cube

crayons

Read Together and Learn

Vocabulary

energy
heat

SCIENCE QUEST Explore energy from the Sun with the Junior Rangers.

What is energy?

When you eat food, you get energy to work and play.

Energy makes things work and change. There are many different forms of energy.

▲ **Gasoline gives cars energy to move.**

◀ **Satellite dishes send energy from space to make pictures on TVs.**

Heat, light, sound, and electricity are some forms of energy.

We use different forms of energy every day.

Quick Lab

Use energy from the Sun to blow up a balloon.

Where do we get energy?

▲ **Windmills turn energy from wind into electricity.**

▲ **Electrical energy can make lights work.**

What is heat?

Heat is energy that makes things warm. We can get heat from burning things like wood, oil, or gas.

People can use this energy to warm their homes.

▲ **Most of the heat energy on Earth comes from the Sun.**

Heat Energy

Read a Photo

What happens when wood burns?

LOG ON *Science in Motion* Watch what heat can do at www.macmillanmh.com

People also use heat to cook.

Rubbing things together is a source of heat, too. You can feel the heat when you rub your hands together.

▲ **Heat popcorn kernels and they will pop!**

✓ What is heat?

Rub your hands together. Friction makes heat. ▶

Think, Talk, and Write

1. **Main Idea and Details.** What are some different ways we get heat?
2. Write about some ways we use energy.

Social Studies Link

How do you use heat where you live?

LOG ON e-Review Summaries and quizzes online at www.macmillanmh.com

Lesson 2

Sound

Look and Wonder

Sounds are made in different ways. How does this man make sounds with his guitar?

Explore

Inquiry Activity

Can you make sound with a rubber band?

What to Do

1. Put a rubber band across a bowl. △ **Be Careful.** Remember to wear safety goggles.
2. **Observe.** Pull the rubber band. Let go. What do you see and hear?
3. Pull it again. Stop the rubber band from moving. What happens?
4. **Draw Conclusions.** What do you think made the sound?

Explore More

5. **Investigate.** Find out if a thicker rubber band makes the same sound.

You need

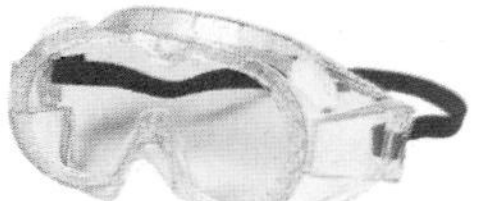

goggles

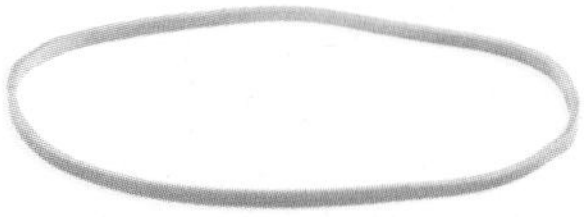

rubber band

plastic bowl

Step 1

Read Together and Learn

Vocabulary

vibrate

pitch

How can you make sound?

You can not see sound, but you can hear it. Sometimes you can even feel it.

Sound is a form of energy. It is made when an object vibrates. **Vibrate** means to move back and forth. When an object stops vibrating, the sound stops, too.

When you hit cymbals together, the metal vibrates, making a sound.▼

FACT When you speak, cords in your throat vibrate.

Different things make different sounds. Sounds can tell us things.

A clock's alarm tells you when to wake up. Fire alarms and car horns can warn you about danger.

✓ **How can sounds help you stay safe?**

Quick Lab

Make an instrument out of different objects.

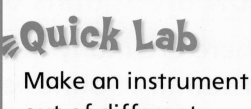

City Sounds

Read a Photo

What sounds would you hear here?

How are sounds different?

There are many different kinds of sounds. Sounds can be loud or soft.

Big vibrations make loud sounds. Small vibrations make soft sounds.

Are these sounds loud or soft?

Some sounds, such as a whistle, are high. Others, like a motorcycle, are low.

A sound's **pitch** is how high or low it is. Fast vibrations make a high pitched sound. Slow vibrations make a low pitched sound.

✓ **What are some soft sounds?**

This coyote's howl is high pitched.

Think, Talk, and Write

1. **Summarize.** How is sound made?
2. Write or draw about how sounds can help you.

Social Studies Link

Listen to sounds in your neighborhood. What makes the sounds you hear?

LOG ON e-Review Summaries and quizzes online at www.macmillanmh.com

Sounds and Safety

There are different kinds of sounds.

Some sounds warn you about danger. They can help you stay safe.

◀ **Fire alarms are very loud. They tell you to move to a safe place.**

▲ **If there is an emergency, call 911. An operator will tell you what you should do and send help.**

Call 911

Some people can not hear. They use their other senses to stay safe.

They can see the flashing lights of an alarm, police car, or ambulance. This warns them of danger.

Smoke alarms beep and flash lights to warn you of danger.

Sirens and flashing lights warn other cars on the road of an emergency. ▼

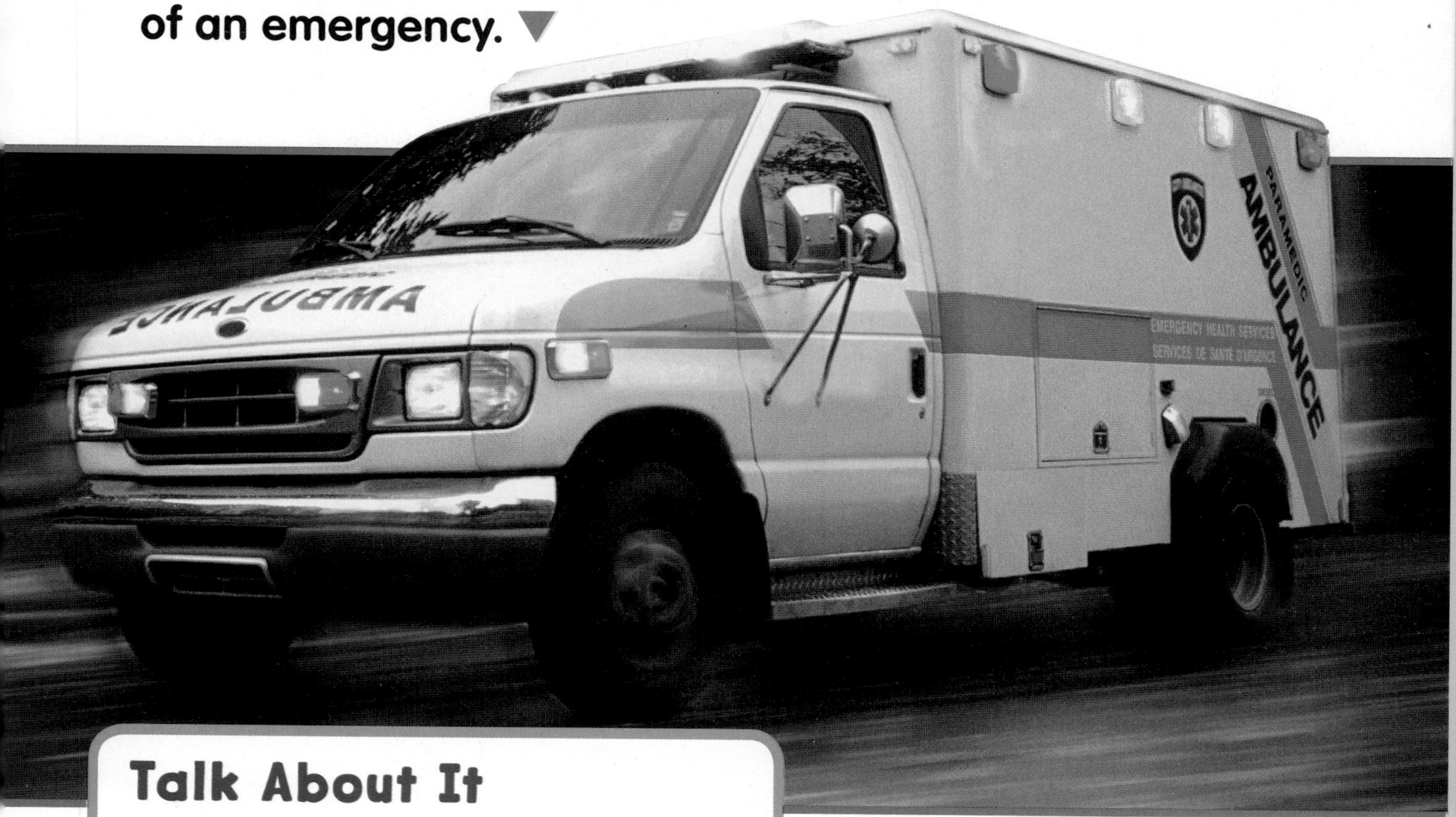

Talk About It

Summarize. Tell how sounds can help you stay safe.

Lesson 3

Light

Barcelona, Spain

Look and Wonder

Stained glass windows are made with different colors of glass. What happens when light shines through the glass?

Explore

Inquiry Activity

What lets light through?

What to Do

1. **Observe.** Look through a cardboard tube. Can you see light in the tube?
2. Cover the end of the tube with aluminum foil. Can you see light now?
3. **Record Data.** Repeat with wax paper and plastic wrap. Record whether or not you can see light.
4. **Draw Conclusions.** Why does light go through some materials and not others?

Explore More

5. **Predict.** What other materials will let light through? Try it.

You need

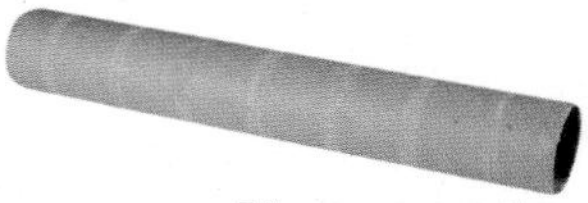

cardboard tube

aluminum foil

wax paper

plastic wrap

rubber band

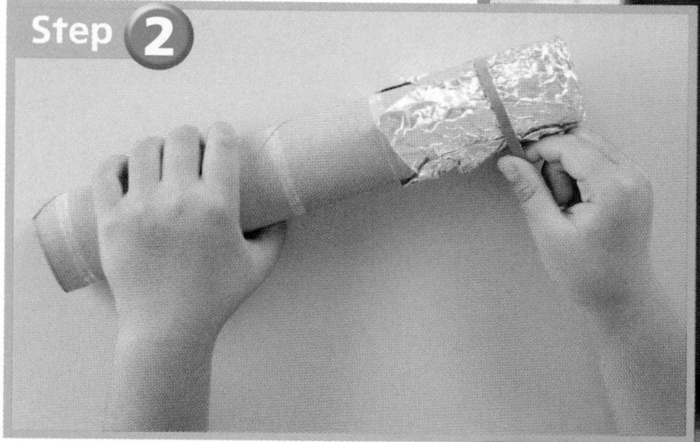

Read Together and Learn

Vocabulary

light

What is light?

Light is a form of energy that lets you see. Different objects let different amounts of light pass through them.

Some objects do not allow any light to go through them.

Light and Sight

This is what the boy sees with regular glasses on.

This is what the boy sees with sunglasses on.

When light is blocked, there is a shadow.

Sometimes your body blocks light. This forms a shadow on the ground.

What are some objects that light can not go through?

▲ This girl has a shadow because light can not shine through her body.

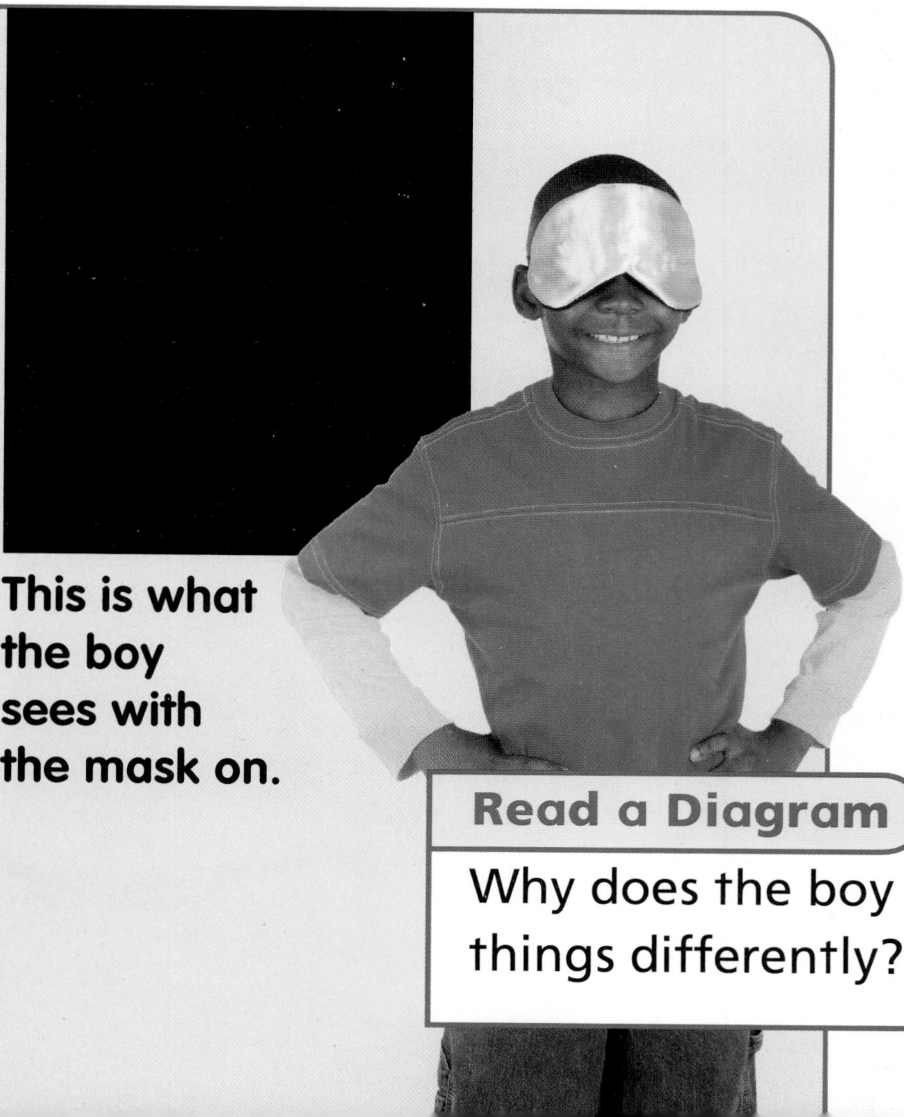

This is what the boy sees with the mask on.

Read a Diagram

Why does the boy see things differently?

What are some sources of light?

Most of the light on Earth comes from the Sun. Stars also make light.

Other lights are made by people. Lamps, streetlights, and flashlights give off light, too.

Quick Lab

Shine light on a mirror. **Record** what you see.

▼ **Streetlights help you see at night.**

▼ **You need light to see your homework.**

Light lets us see things. When light hits an object it bounces off the object.

Then the light goes into your eyes. This lets you see the object.

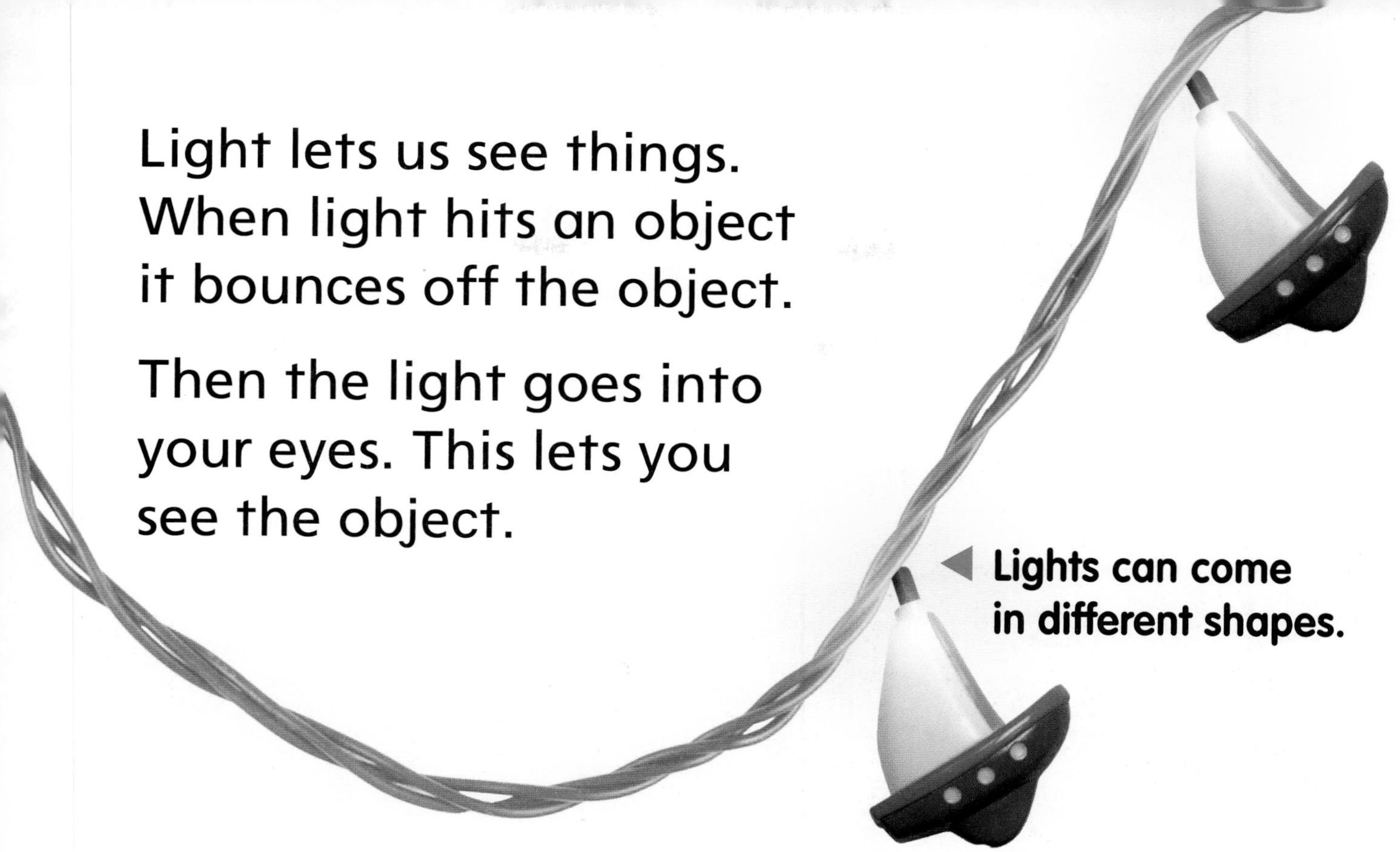

◀ Lights can come in different shapes.

✔ **Why is light important?**

Think, Talk, and Write

1. Put Things in Order. How does your body make a shadow?
2. Write or draw about how you use light every day.

Health Link

Light allows us to see. Find out how to take care of your eyes.

LOG ON e-Review Summaries and quizzes online at www.macmillanmh.com

Turn On the Lights

We get light from many places. We can get light from the Sun or a lamp. Light helps us see things.

Write About It

Write a story about the different kinds of light in this picture. How does the man use the lights?

Remember
Use details to describe a picture.

LOG ON e-Journal Write about it online at www.macmillanmh.com

Stained Glass

Stained glass windows are made with many pieces of colored glass. When sunlight shines through stained glass, you can see different colors of light.

Sort the Shapes

What shapes do you see in the stained glass window above? How many circles do you see? How many rectangles do you see?

Remember
Use tally marks to keep track of the shapes you counted.

Lesson 4

Electricity

California

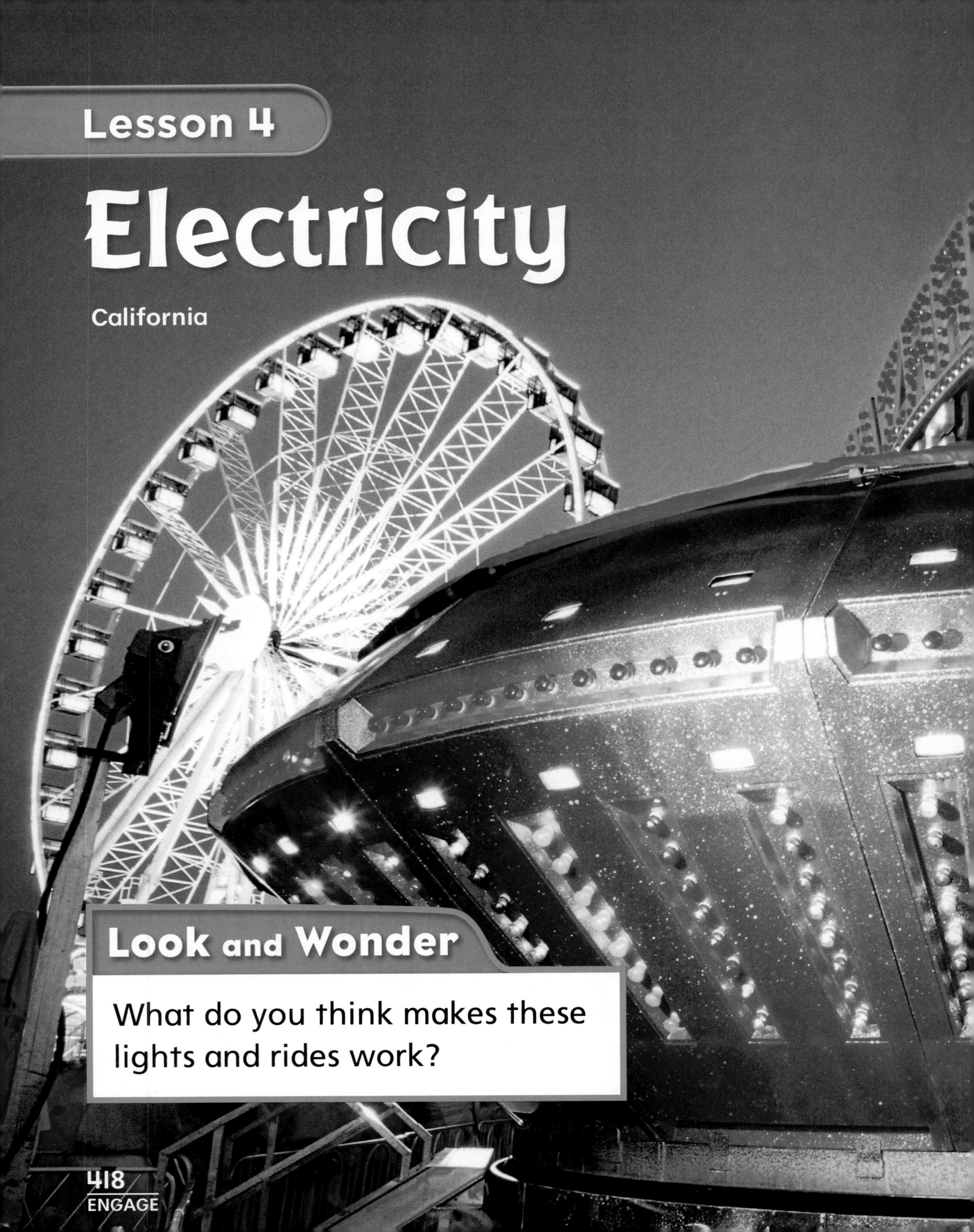

Look and Wonder

What do you think makes these lights and rides work?

Explore

Inquiry Activity

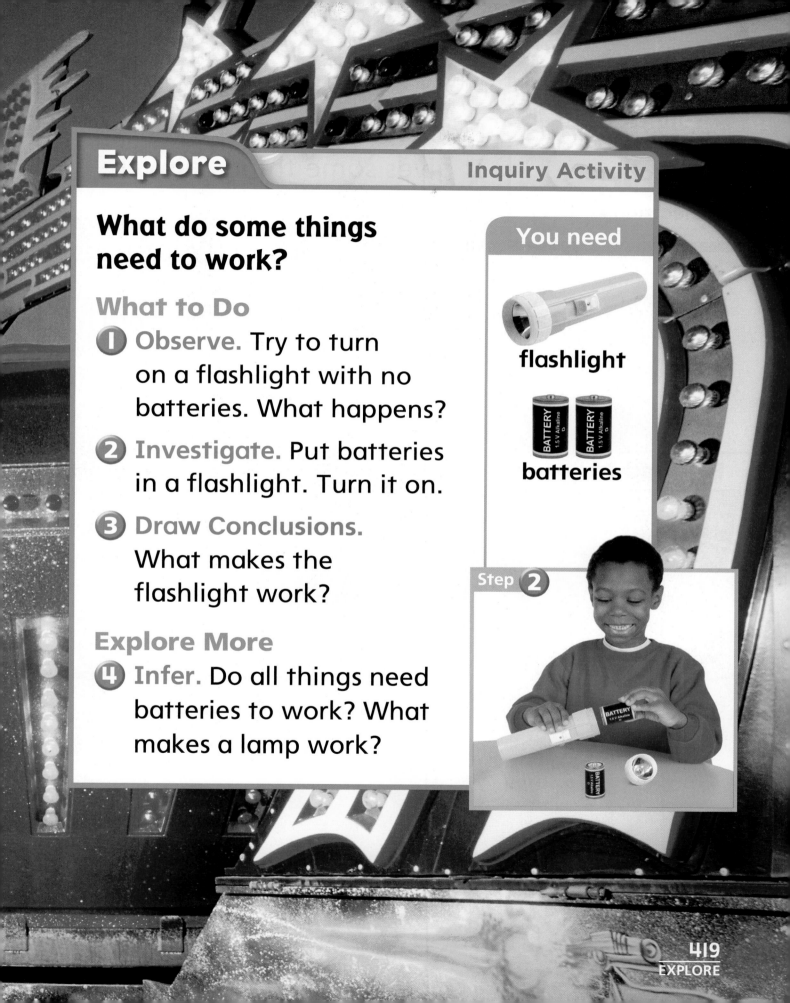

What do some things need to work?

What to Do

1. **Observe.** Try to turn on a flashlight with no batteries. What happens?
2. **Investigate.** Put batteries in a flashlight. Turn it on.
3. **Draw Conclusions.** What makes the flashlight work?

Explore More

4. **Infer.** Do all things need batteries to work? What makes a lamp work?

You need

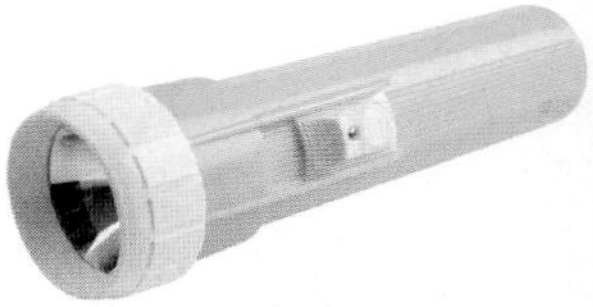

flashlight

batteries

Step 2

Read Together and Learn

Vocabulary

electricity

How do you use electricity?

Electricity is a form of energy. It gives some things power to work.

Wires carry electricity into your school and home. You can also get electricity from batteries.

Plugged In

Without electricity, many things we use every day would not work.

Some electrical things can be dangerous. Never use electricity near water.

✔ **What things do you use that need electricity to work?**

Quick Lab

Find things in your school that use electricity.

Read a Diagram

Which objects in this room need electricity to work?

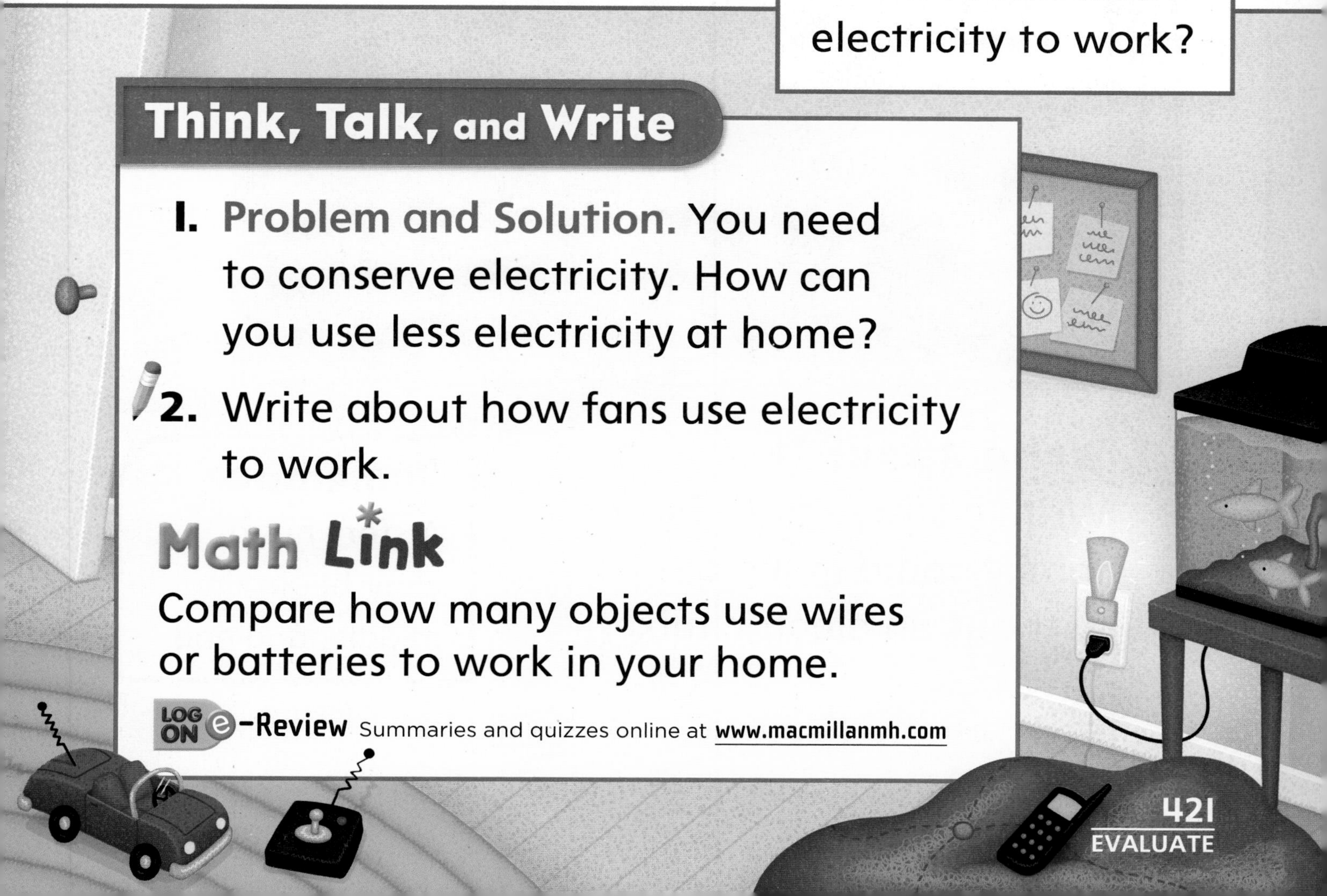

Think, Talk, and Write

1. **Problem and Solution.** You need to conserve electricity. How can you use less electricity at home?
2. Write about how fans use electricity to work.

Math Link

Compare how many objects use wires or batteries to work in your home.

LOG ON e-Review Summaries and quizzes online at www.macmillanmh.com

Electricity at Home

Many kitchen appliances, like mixers, need electricity to work. They can make work easier. How do you think the electric mixer makes cooking easier for this family?

Write About It

Write a story about how this family could make dinner without an electric mixer.

Remember

A story has a clear beginning, middle, and end.

LOG ON e-Journal Write about it online at www.macmillanmh.com

You need light in order to see.
You need electricity for the TV.

CHAPTER 12 Review

Vocabulary

Use each word once to complete the sentences.

electricity
heat
pitch
vibrates

1. The highness or lowness of a sound is its ____.

2. You need ____ to make a computer work.

3. A sound is made when an object ____.

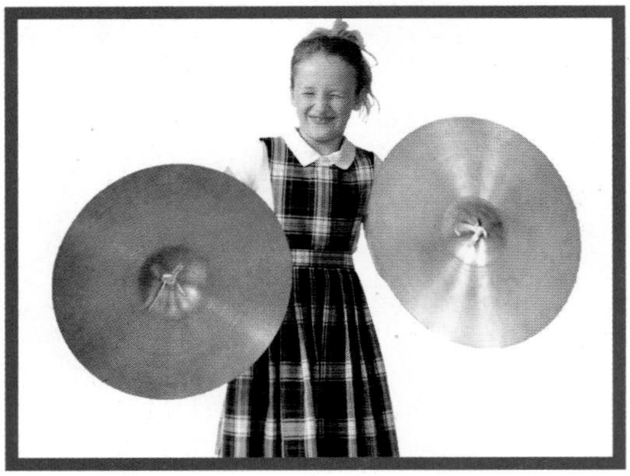

4. When you rub your hands together, you can feel ____.

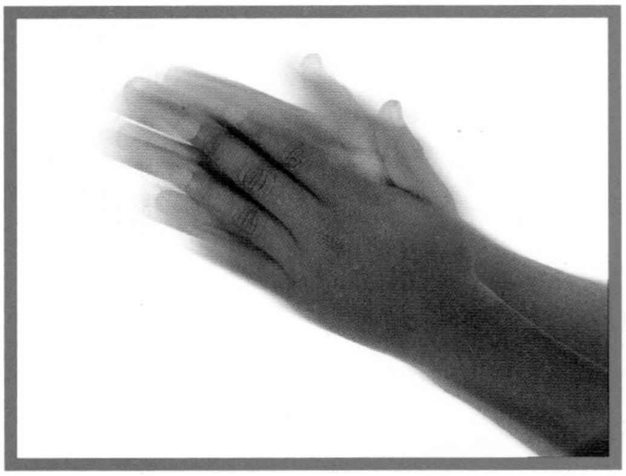

Answer the questions below.

5. How could you make different sounds on a guitar?

6. **Draw Conclusions.** Talk about the forms of energy in these pictures.

7. **Summarize.** Where can we get heat?

8. What makes shadows on the ground?

The Big Idea

9. What is energy?

Musician

Do you like to sing or play an instrument? Do you like sound? You could become a musician. A musician's job is to make different sounds.

musician

Musicians need to study and practice hard. Musicians have to know all about fast, slow, high, and low sounds. Jazz, classical, and rock are kinds of music. What kind of music do you like?

More Careers to Think About

sound engineer

instrument maker

LOG ON e-Careers at www.macmillanmh.com

Reference

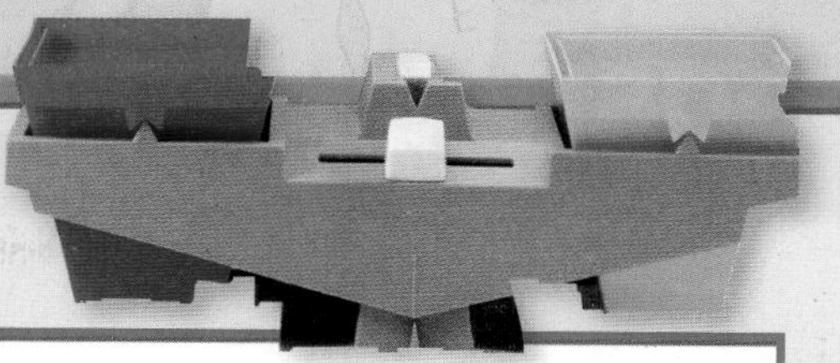

Science Handbook

Health Handbook

Measurements

Nonstandard

You can use objects to measure the length of some solids. Line up objects and count them. Use objects that are alike. They must be the same size.

▲ **This string is about 8 paper clips long.**

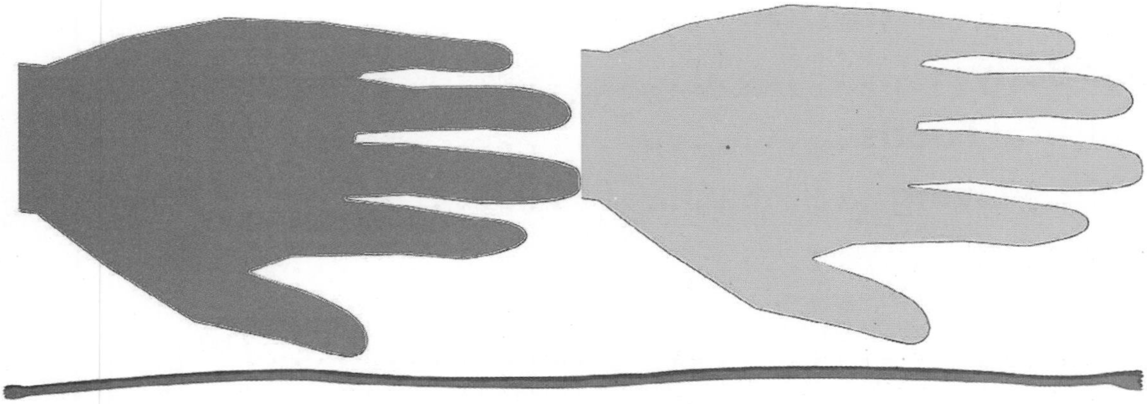

▲ **This string is about 2 hands long.**

Try It

Measure a solid in your classroom. Tell how you did it.

Standard

You can also use a ruler to measure the length of some solids. You can measure in a unit called **centimeters**.

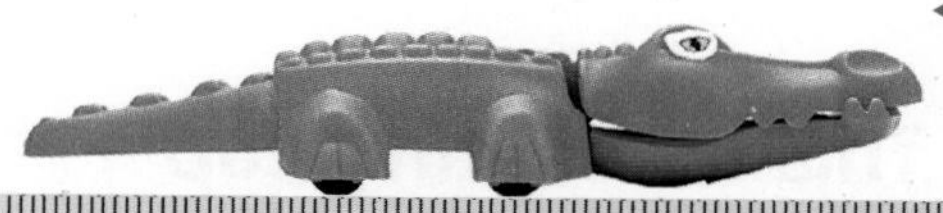

◀ **This toy is about 8 centimeters long. This is written as 8 cm.**

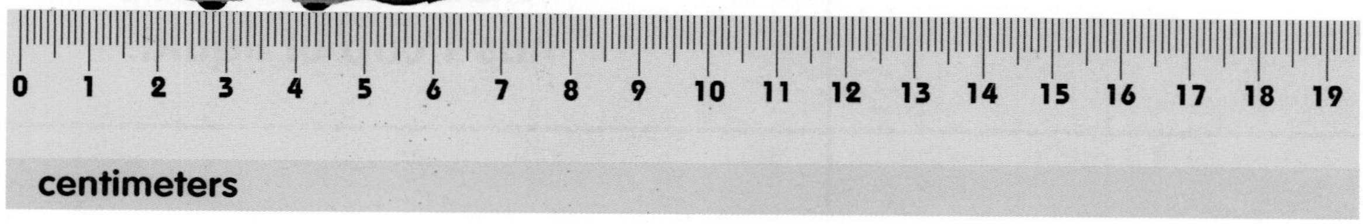

You can also use a ruler to measure in a unit called **inches**. One inch is longer than 1 centimeter.

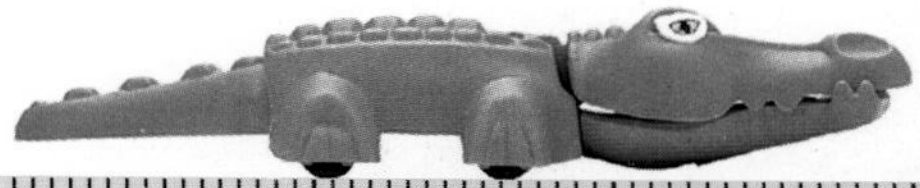

◀ **This toy is about 3 inches long. This is written as 3 in.**

Try It

Estimate the length of this toy car. Then find its exact length.

Measurements

Volume

You can measure the volume of a liquid with a **measuring cup**. Volume is the amount of space a liquid takes up.

▲ **This measuring cup has 1 cup of liquid.**

Mass

You can measure mass with a **balance**. The side that has the object with more mass will go down.

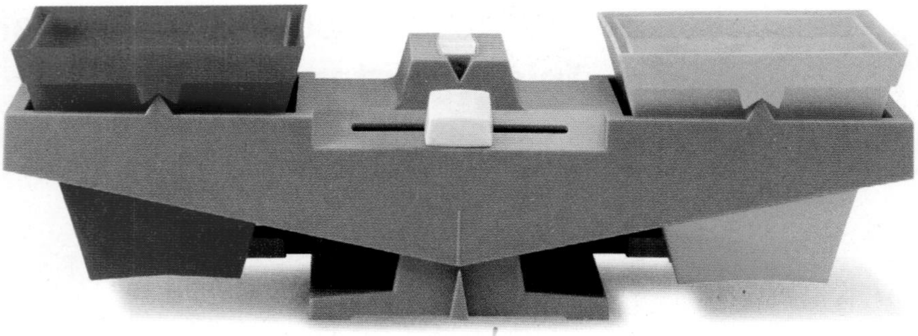

▲ **Before you compare the mass of two objects, be sure the arrow points to the line.**

Try It

Place two objects on a balance. Which has more mass?

Time

You can measure time with a **clock**. A clock measures in units called hours, minutes, and seconds. There are 60 minutes in 1 hour.

Temperature

You can measure temperature with a **thermometer**. Thermometers measure in units called degrees.

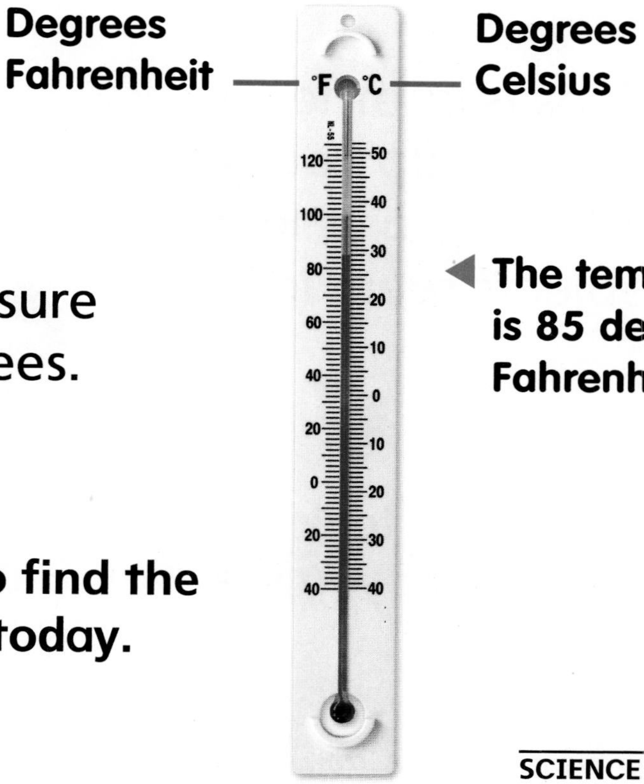

◀ **The temperature is 85 degrees Fahrenheit.**

Try It

Use a thermometer to find the temperature outside today.

Science Tools

Computer

A computer is a tool that can help you get information. You can use the Internet to connect to other computers around the world.

When you use a computer, make sure an adult knows what you are working on.

Hand Lens

A hand lens is another tool that can help you get information. A hand lens makes objects seem larger.

Try It

Use a hand lens to look at an object. Draw what you see.

Graphs

Bar Graphs

Bar graphs organize data. The title of the graph tells you what the data is about. The shaded bars tell you how much of each thing there is.

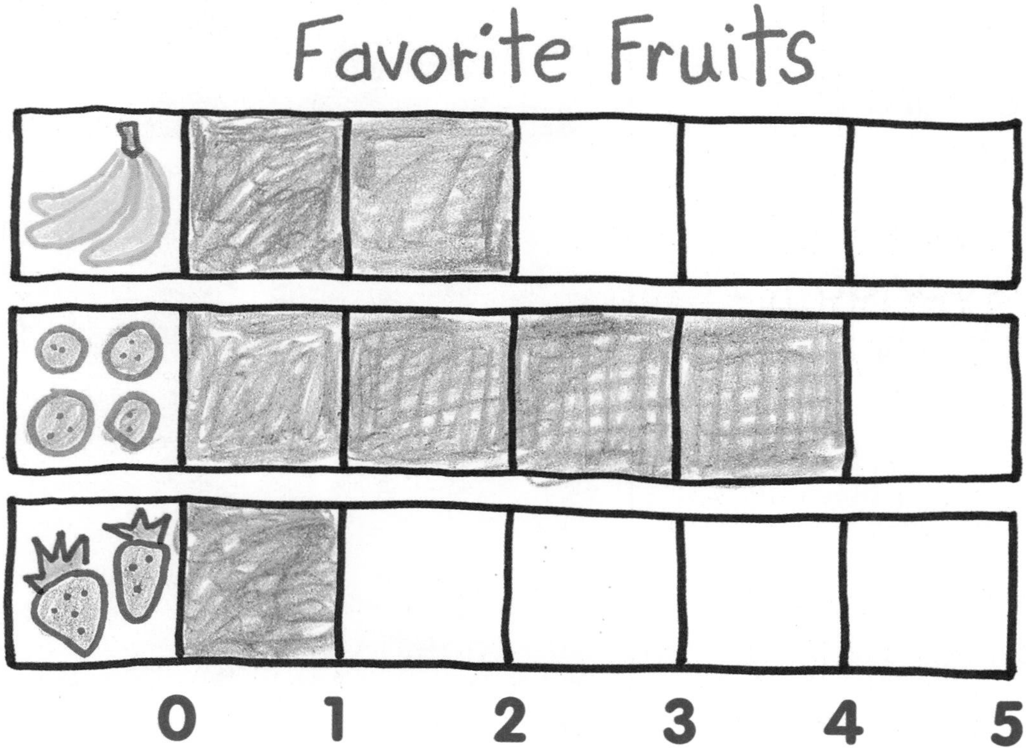

Try It

Make a bar graph that shows your classmates' favorite fruits.

Your Body

Skeletal System

Your body has many parts. All your parts work together to help you live.

Bones are hard body parts inside your body. They help you stand straight. Your bones give your body its shape.

Skeletal System

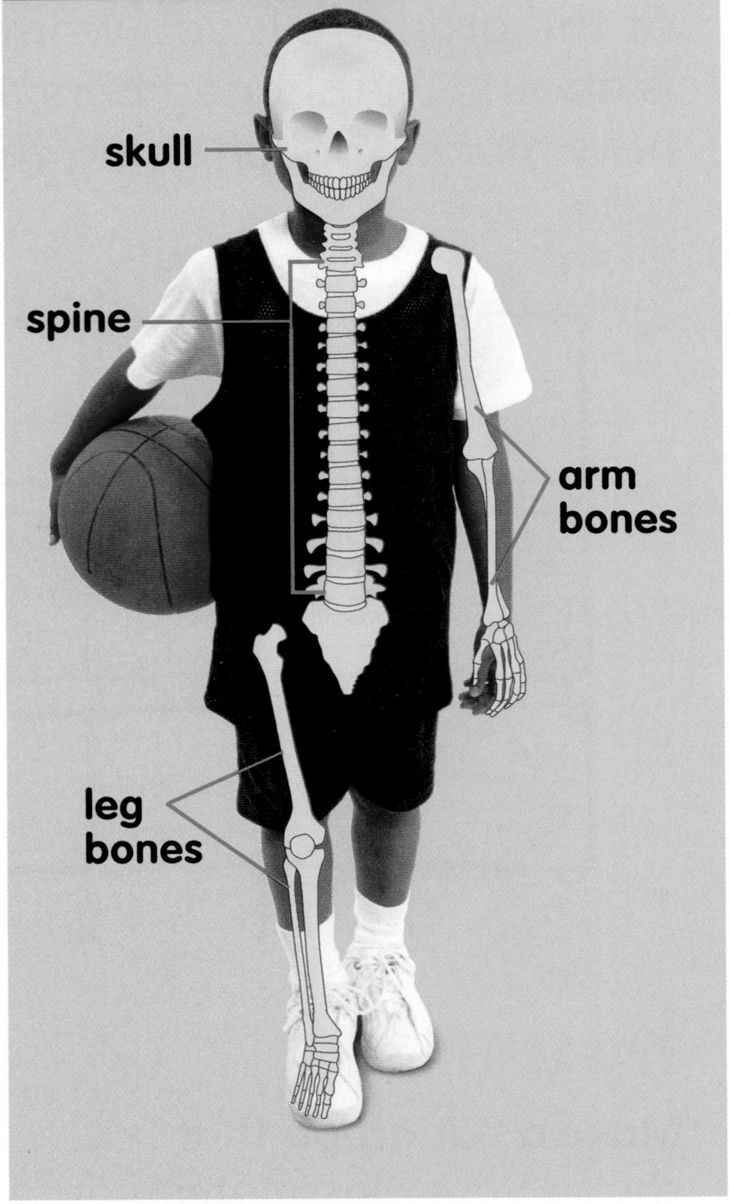

Try It

How many bones do you think there are in your arms? Count them.

Muscular System

Muscles are body parts that help you move. They are inside your body.

Muscles get stronger when you exercise them.

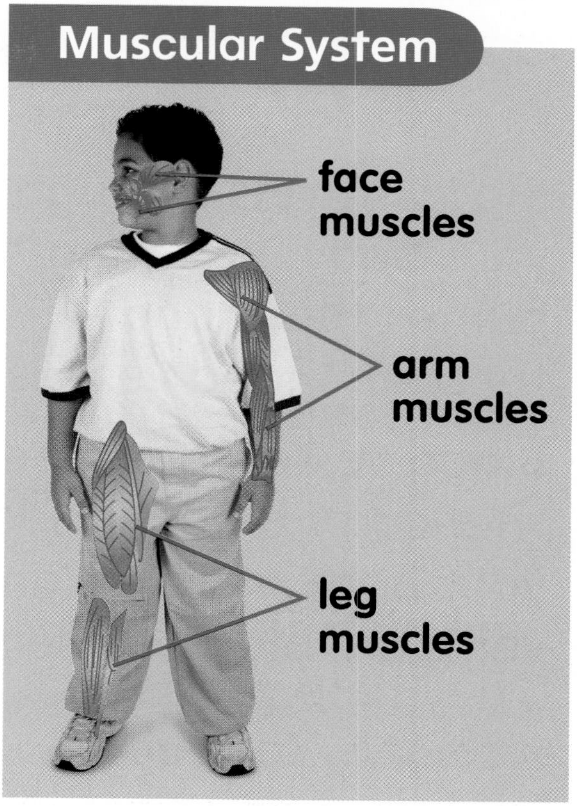

Nervous System

Your brain sends messages all around your body. The messages travel along tiny body parts called nerves.

These messages tell your body parts to move. They can also alert you of danger.

Try It

Jump up and down in place. Which muscles did you use?

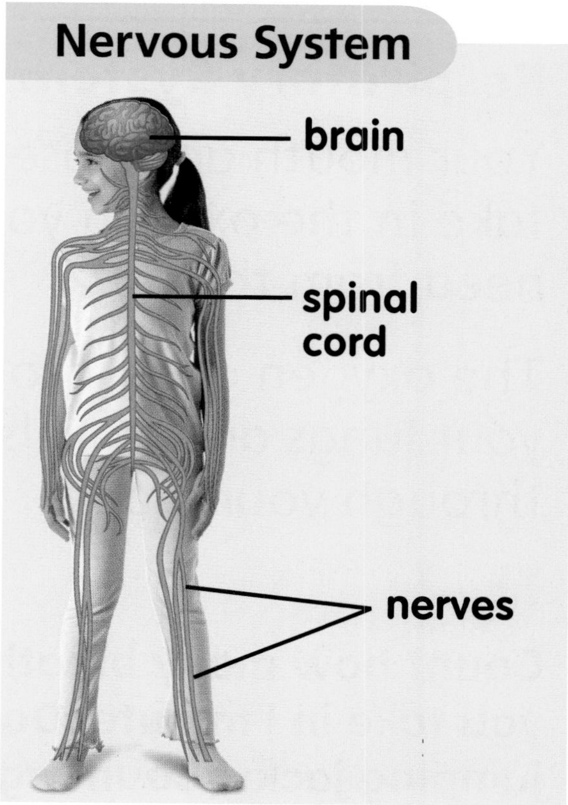

Your Body

Circulatory System

Blood travels through your body. Your heart pumps this blood through blood vessels.

Blood vessels are tubes that carry blood inside your body. Arteries and veins are blood vessels.

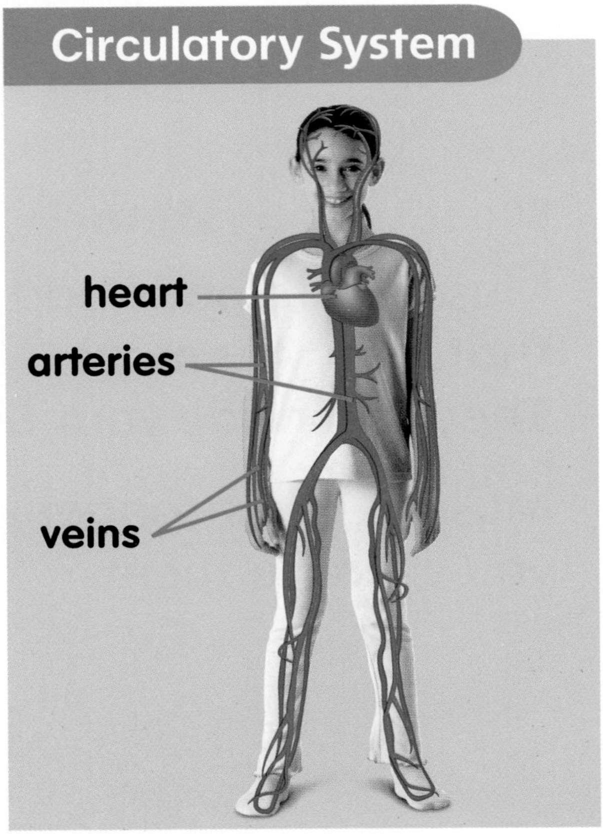

Respiratory System

Your mouth and nose take in the oxygen you need from the air.

The oxygen goes into your lungs and travels through your blood.

Try It

Count how many breaths you take in 1 minute. Do ten jumping jacks. Count again.

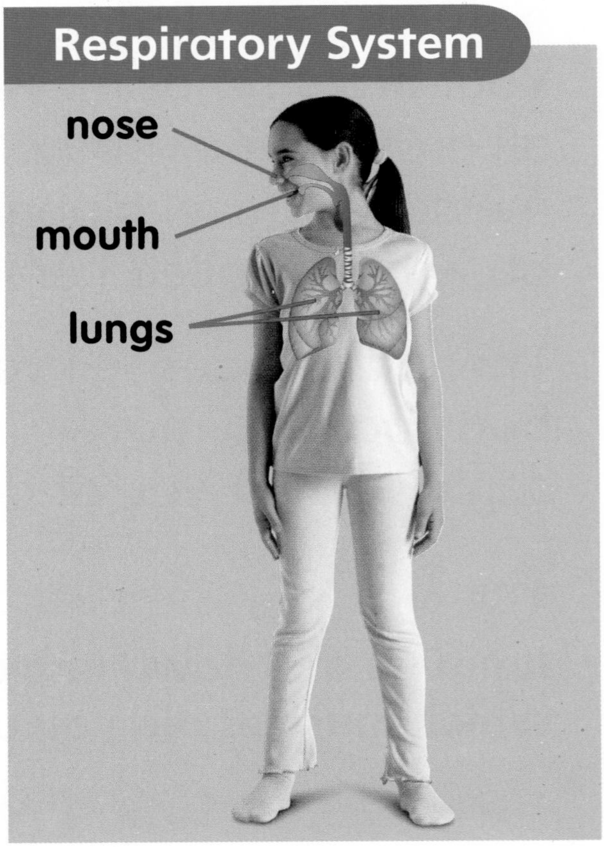

Digestive System

When you eat, your body uses food for energy.

Food enters your body through your mouth. Your stomach and intestines help break down the food in your body. This helps your body get nutrients.

Digestive System

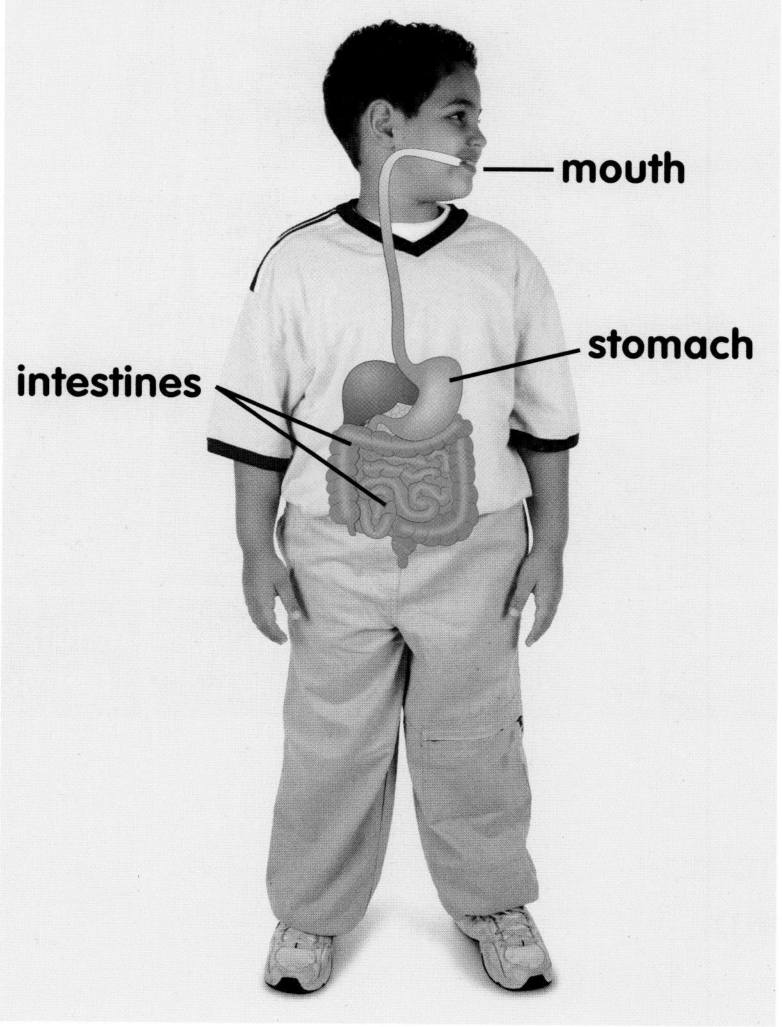

Try It

Name some body parts that help you eat food.

Healthful Foods

MyPyramid

MyPyramid is a guide for healthful eating. A healthful meal contains foods from the five food groups. A food group is a group of foods that are alike.

Eat more foods from the largest slice of the pyramid. Eat less from the smallest slice.

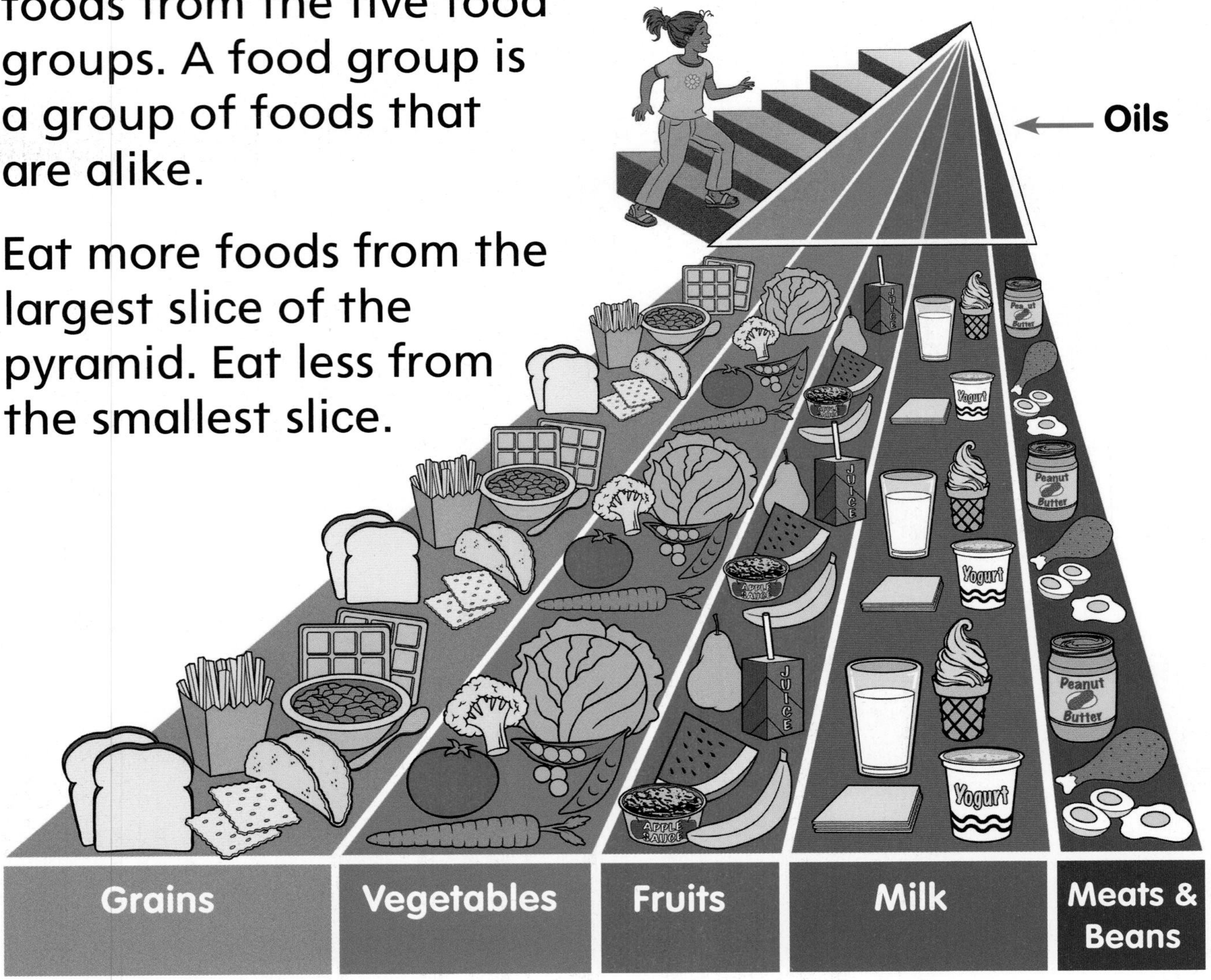

Try It

Plan a healthful meal. Include one food from each group.

Healthful Foods

Foods can come from plants or animals. Grains, fruits, and vegetables give you energy to work and play. Foods in the milk group keep your bones strong. Meats, fish, and beans help make your muscles strong.

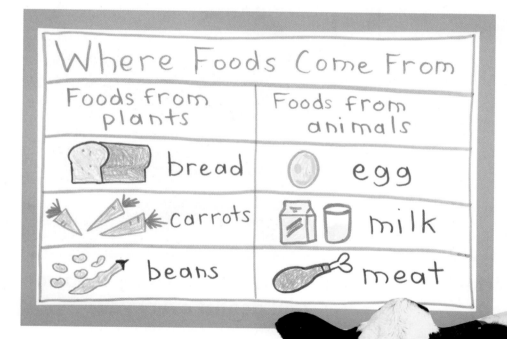

Where Foods Come From	
Foods from plants	Foods from animals
bread	egg
carrots	milk
beans	meat

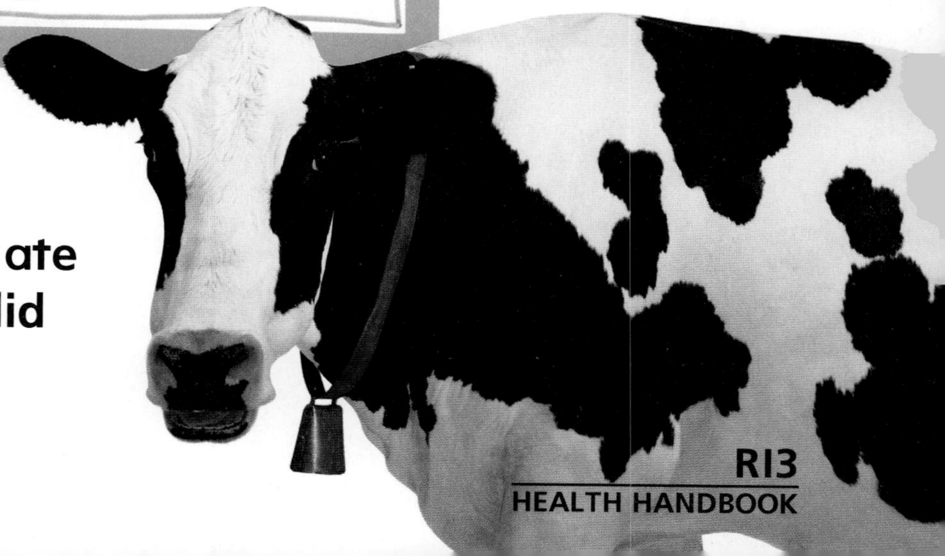

Try It

Write down what you ate for breakfast. Where did the foods come from?

Healthy Living

Stay Healthy

Be active every day. Exercise keeps your heart and lungs healthy.

Doctors and dentists can help you stay healthy as you grow.

▲ **Exercise is important for a healthy body.**

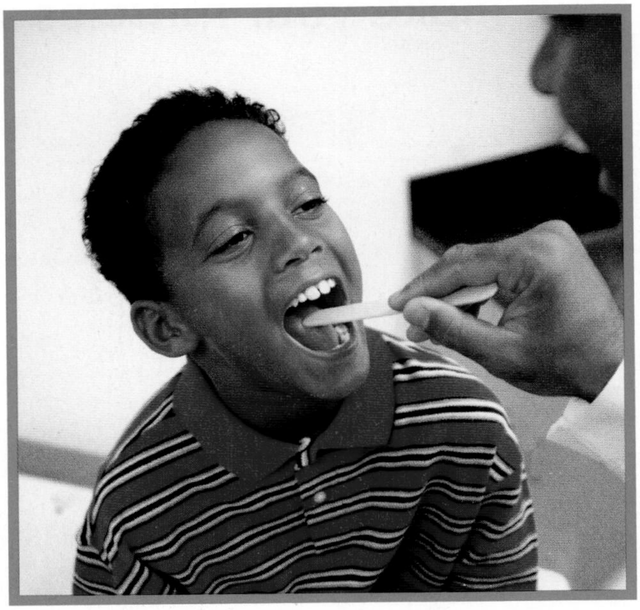

▲ **Get a checkup from a doctor and dentist every year.**

Try It

Record how many times you exercise in one week.

Take Care of Your Body

Tobacco and alcohol harm you. Tobacco smoke can make it hard to breathe. Alcohol slows down your mind and body.

▲ **Only take medicines that your parent or doctor gives you.**

Here are some ways to take care of your body. ▼

Take Care of Yourself
Take a bath.
Brush and floss your teeth every day.
Stand up straight.
Get plenty of sleep.

Try It

Make a poster about being drug free. Share it with your school.

Safety

Safety Indoors

To stay safe indoors, do not touch dangerous things. Tell an adult about them right away. Never taste anything without permission.

In case of a fire, get out fast. If your clothes catch fire, remember to stop, drop, and roll.

▲ **Do not touch these things.**

Try It

Practice stop, drop, and roll. Teach it to a friend.

Safety Outdoors

Be safe outdoors. Follow these rules.

▲ **Wear a helmet.**

▲ **Cross at a crosswalk.**

▲ **Wear your seat belt.**

▲ **Follow game rules.**

Try It

Choose one of the rules. Make a poster showing the safety rule.

Get Along

Work and play well with others. Show others respect and care. Be fair and take turns when playing with one another.

Try It

Make a friendship badge. Give your badge to someone who is being a good friend.

Glossary

A

adaptation A body part or behavior that helps an animal survive. (page 129) A giraffe's long neck is an adaptation.

amphibian An animal that lives on land and in water. (page 91) A frog is an amphibian.

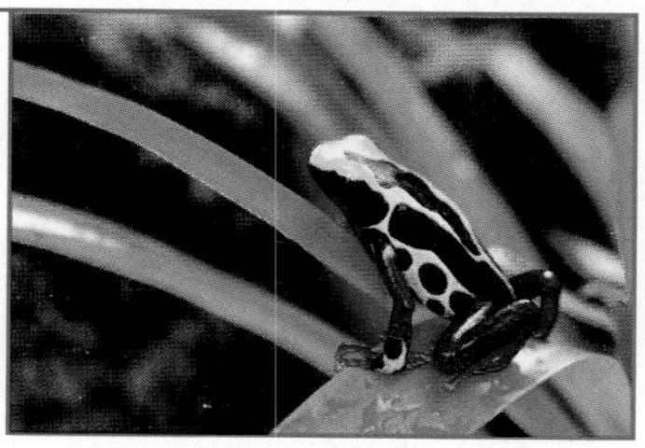

Arctic An icy and cold place near the North Pole. (page 70) When Arctic snow melts, small flowers grow.

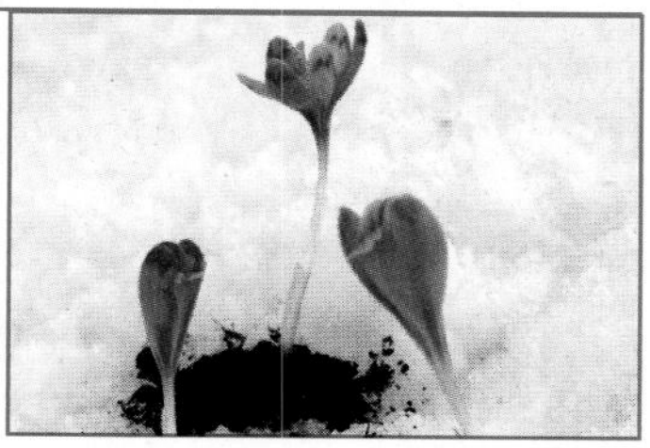

B

balance A tool used to measure mass. (page 302) The side of a balance with more mass will go down.

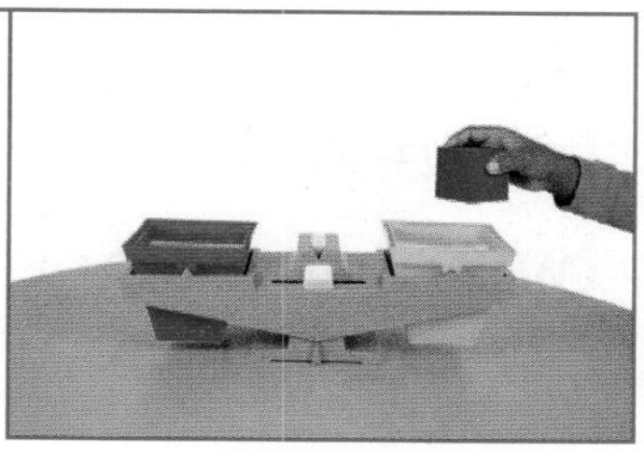

bird An animal that has two legs, two wings, and feathers. (page 89) A duck is a bird.

LOG ON e-Glossary at www.macmillanmh.com

burn A way of changing matter using heat. (page 330) When you burn paper, it changes to ash.

C

carnivore An animal that eats other animals. (page 106) A tiger is a carnivore.

cloud Tiny drops of water and bits of ice that collect in the sky. (page 237) Rain or snow can fall from a cloud.

conserve To save, keep, or protect. (page 210) You can conserve resources by not wasting them.

continent A large piece of land on Earth. (page 164) There are seven continents on Earth.

D

desert A dry place. (page 68) Cactus plants can live in the desert.

dissolve To completely mix into a liquid. (page 336) Drink mix will dissolve in water.

E

electricity A form of energy that gives some things the power to work. (page 420) Many things in your home need electricity to work.

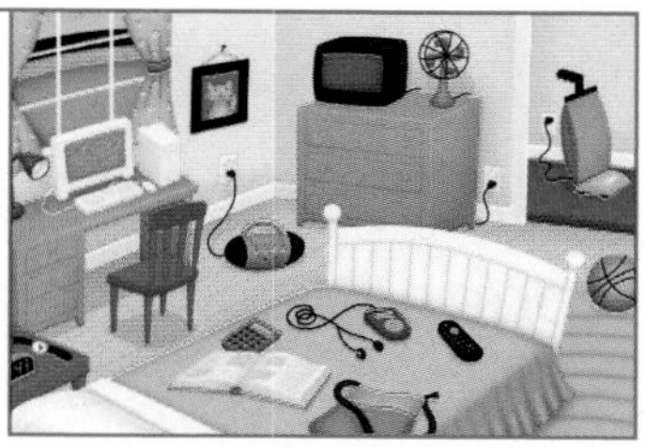

energy A force that makes things work or change. (page 398) Gasoline gives cars the energy to move.

erosion When rock and soil are moved by wind or water to a new place. (page 182) Erosion slowly changes the shape of land.

evaporate To change from a liquid to a gas. (page 344) Heat from the Sun made the water in this pond evaporate.

extinct When all of one kind of plant or animal dies. (page 147) The woolly mammoth is an extinct animal.

F

fall The season after summer. (page 250) Some leaves change colors in fall.

fish An animal that lives in water and has gills and fins. (page 92) Fish use gills to breathe in water.

flower A part of a plant that makes seeds. (page 54) Flowers come in many shapes and colors.

food chain The order in which living things get food in a habitat. (page 144) All animals are part of a food chain.

force A push or a pull that makes an object move. (page 368) It takes force, like a push, to move a ball.

forest A place where there are many tall trees. (page 130) Many plants and animals live in the forest.

freeze To change from a liquid to a solid. (page 342) Water will freeze if it gets very cold.

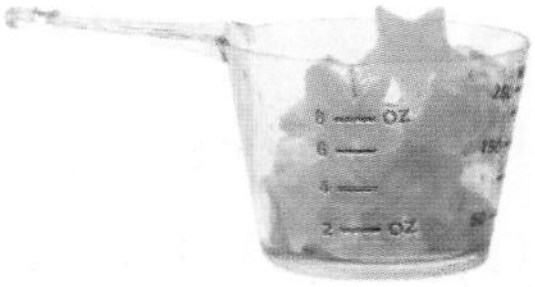

friction A force that slows things down. (page 371) If you drag a rubber stopper on the ground, friction makes you stop.

fruit The plant part that holds the seeds. (page 55) The peach fruit has a seed inside.

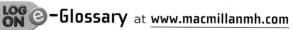

G

gas A state of matter that does not have its own shape. (page 318) Gas gives balloons their shape.

gills The part of a fish that takes in oxygen from water. (page 99) A fish uses its gills to breathe in water.

grassland A large open place with a lot of grass. (page 128) Prairie dogs live on a grassland.

gravity A force that pulls things toward Earth. (page 369) Gravity keeps us from staying up in the air.

H

habitat A place where plants and animals live. (page 128) A forest is a habitat for many plants and animals.

hatch A baby animal breaking out of an egg. (page 112) Birds hatch from eggs.

heat A form of energy that makes things warm. (page 400) Heat can make popcorn pop.

herbivore An animal that eats plants. (page 105) A rabbit is a herbivore.

I

insect An animal with three body parts and six legs. (page 93) An ant is an insect.

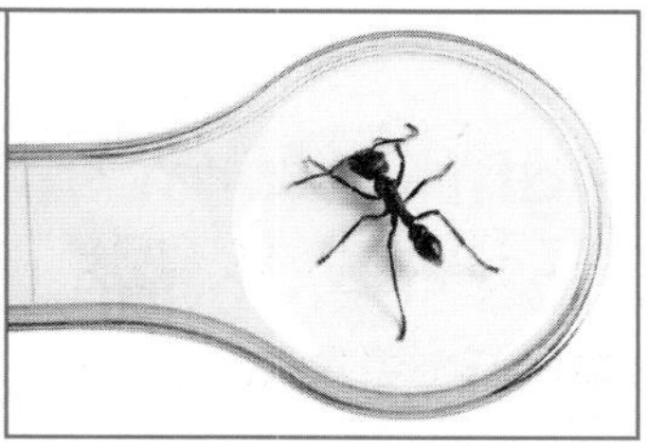

L

lake Water that has land all around it. (page 134) A lake can be a home for many plants and animals.

leaves Plant parts that use sunlight and air to make food. (page 32) Leaves come in different shapes and sizes.

lever A bar that balances on a point and moves like a seesaw. (page 376) An oar is a lever.

life cycle How a living thing grows, lives, and dies. (page 60) The life cycle of a bean plant starts with a seed.

light A form of energy that lets you see. (page 412) Light from a flashlight can help you see in the dark.

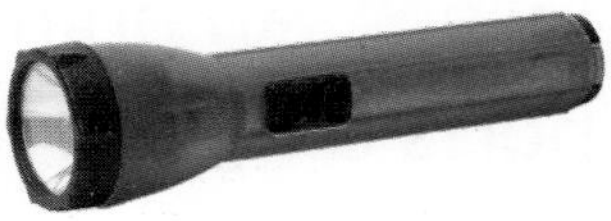

liquid A state of matter that flows and takes the shape of its container. (page 316) Milk is a liquid.

living Something that grows, changes, and needs food, air, and water to survive. (page 24) This girl is a living thing.

lungs Body parts used to breathe air. (page 99) Birds use lungs to breathe.

M

magnet Something that can pull, or attract, some objects with metal in them. (page 382) A magnet can attract metal paper clips.

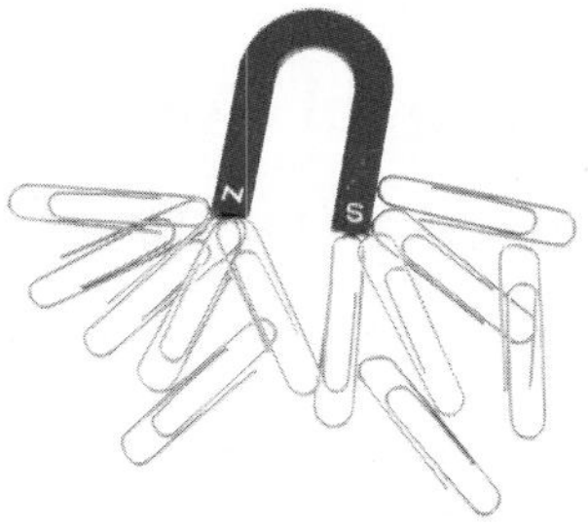

mammal An animal with hair or fur. (page 88) Most mammals give birth to live young.

mass The amount of matter in an object. (page 302) A metal bird has more mass than a sponge bird.

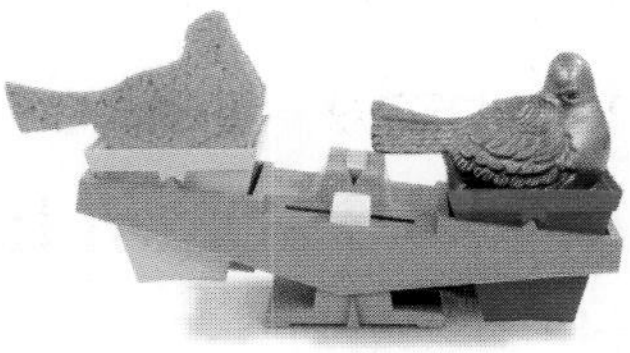

matter What all things are made of. (page 301) A kite is made of matter.

melt To change from a solid to a liquid. (page 343) Ice cubes can melt and become water.

mineral A nonliving thing from the earth. (page 173) All rocks are made of minerals.

mixture Two or more different things put together. (page 334) A fruit salad is a mixture of different fruits.

Moon A ball of rock that moves around Earth. (page 280) The Moon does not make its own light.

motion A change in an object's position. (page 364) The airplane is in motion.

mountain Land that is very high. (page 168) A mountain is the highest type of land.

N

natural resource Something that comes from Earth that people use. (page 196) Rocks are a natural resource.

nonliving Something that does not grow and change, or need food, air, or water to survive. (page 25) A rock is a nonliving thing.

nutrient Something that living things need to grow. (page 26) Plant roots can get nutrients from soil.

O

ocean Salty water that is very large and deep. (page 136) Whales live in the ocean.

P

phases The different Moon shapes we see each month. (page 281) A crescent moon is one of the Moon's phases.

LOG ON e-Glossary at www.macmillanmh.com

pitch The lowness or highness of a sound. (page 407) The sound of a siren has a high pitch.

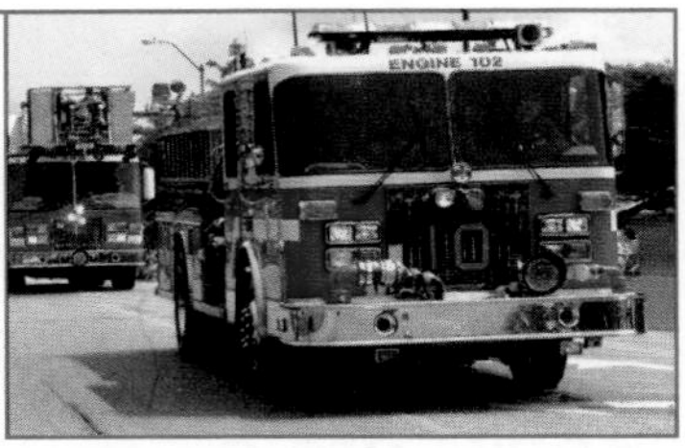

plain Flat land that spreads out a long way. (page 169) A plain is wide and flat.

planet A very large object that moves around the Sun. (page 282) Saturn is a planet.

poles The places where a magnet's pull is strongest. (page 384) A magnet has a North pole and a South pole.

pollution Harmful things in the air, land, or water. (page 204) Water pollution can harm animals.

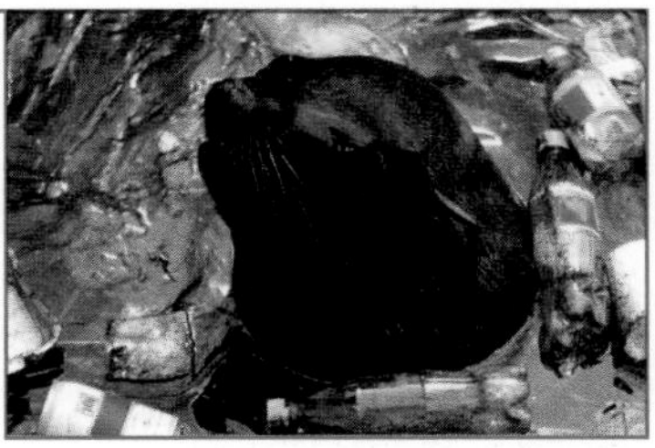

position The place where something is located. (page 362) You can find something if you know its position.

property How something looks, feels, smells, tastes, or sounds. (page 300) Color is a property of the puppets.

pull A force that moves something closer to you. (page 368) The boy pulls the bag of basketballs.

pulley A rope that moves over a wheel. (page 375) A pulley can help raise a flag to the top of a flagpole.

push A force that moves something away from you. (page 368) The girl pushes the basketball when she throws it.

R

rain forest A hot, wet place. (page 69) A rain forest has many green plants.

rain gauge A tool that measures how much rain falls. (page 232) A rain gauge is a weather tool.

LOG ON e-Glossary at www.macmillanmh.com

ramp A slanted surface that you can use to move things up or down. (page 377) A ramp makes it easier to go up to a higher place.

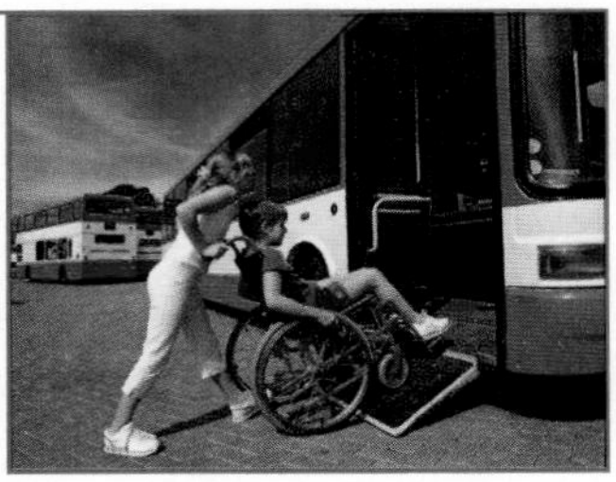

recycle To make a new thing from an old thing. (page 213) You can recycle paper, plastic, and glass.

reduce To use less of something. (page 212) You can reduce how much water you use by turning the water off when brushing your teeth.

repel To push away. (page 385) Alike poles on magnets will repel each other.

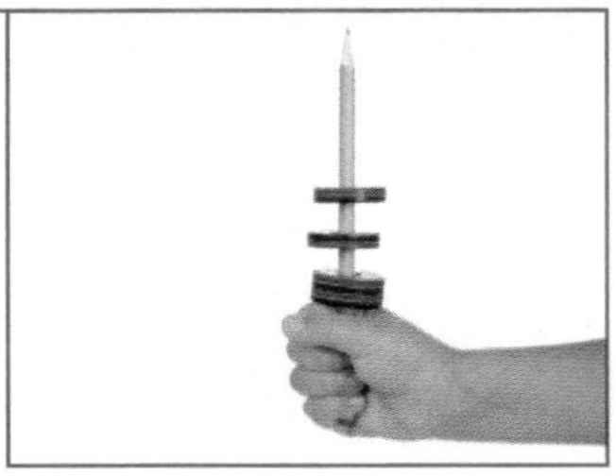

reptile An animal that has dry skin covered with scales. (page 90) A snake is a reptile.

reuse To use something again. (page 210) You can reuse cans and bottles.

river Fresh water that moves. (page 167) A river may flow into an ocean.

root Plant part that keeps the plant in the ground. (page 32) Roots hold a plant in the ground.

rotate To spin. (page 272) Earth turns, or rotates, every 24 hours.

S

season A time of year. (page 242) Fall, winter, spring, and summer are the four seasons.

seed A part of a plant that can grow into a new plant. (page 54) A seed inside a peach can grow into a peach tree.

seedling A young plant. (page 60) A young bean plant is a seedling.

shelter A place where animals can live and be safe. (page 97) These raccoons find shelter in a log.

simple machine A tool that can make it easier to move things. (page 374) A rake is a simple machine.

soil The top layer of Earth. (page 174) Soil is a mixture of tiny bits of rock, air, water, dead plants, and dead animals.

solid A state of matter that has a shape of its own. (page 308) A block is a solid.

speed How fast or slow something moves. (page 365) A rocket ship can move at a fast speed.

spring The season after winter. (page 242) Many baby animals are born in spring.

star An object in the sky that makes its own light. (page 266) We can see many stars in the night sky.

stem The part of a plant that holds up the plant. (page 32) The stem holds up the flower.

summer The season after spring. (page 244) Lemonade can cool you off in the hot summer.

Sun The star closest to Earth. (page 267) The Sun gives light and heat to Earth.

tadpole A young frog. (page 114) A tadpole grows into an adult frog.

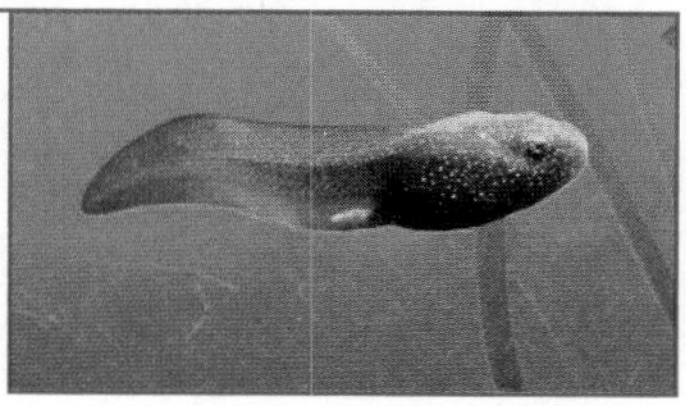

LOG ON e-Glossary at www.macmillanmh.com

T

temperature How hot or cold something is. (page 231) In winter, the temperature can be very cold.

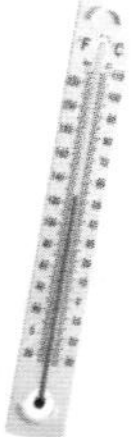

thermometer A tool that measures temperature. (page 232) The thermometer shows a temperature of 65 degrees Fahrenheit.

trunk The thick stem of a tree. (page 39) A trunk helps protect a tree from weather and animals.

V

valley Low land between mountains. (page 168) The valley is flat.

vibrate To move back and forth quickly. (page 404) Sound is made when something vibrates.

W

water vapor Water that goes up into the air as a gas and is too small to see. (page 236) You can not see water vapor.

weather What the sky and air are like each day. (page 230) The weather is rainy today.

weathering When water changes the shape and size of rocks. (page 180) Weathering can make rocks crack.

wind vane A tool that shows the direction of the wind. (page 232) A wind vane is a weather tool.

winter The season after fall. (page 252) It can snow in winter.

Science Skills

classify To group things by how they are alike. (chapter 2) You can classify these animals as birds because they have feathers.

---

communicate To write, draw, or tell your ideas. (chapter 4) You can use a word web to communicate your ideas to others.

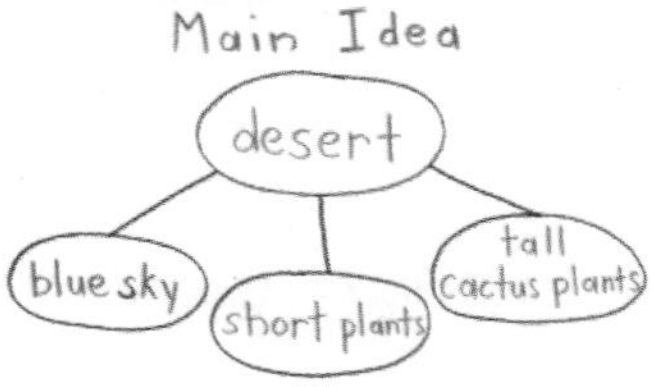

---

compare To observe how things are alike or different. (chapter 3) You can use a Venn diagram to compare two things.

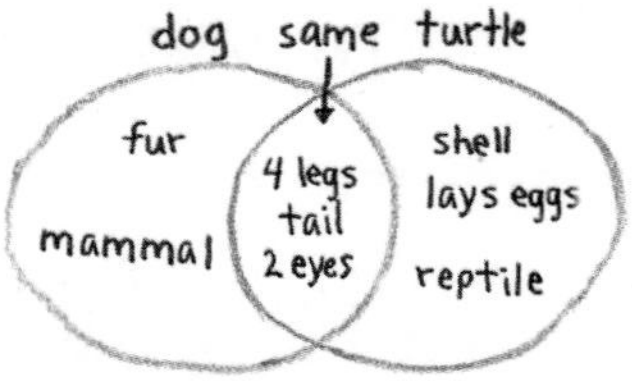

---

draw conclusions To use what you observe to explain what happens. (chapter 12) You can draw the conclusion that the bird was hungry because it ate all the sunflower seeds.

---

infer To use what you know to figure something out. (chapter 11) Since there are not many plants to eat in winter, you can infer that the bird does not have a lot of food.

---

investigate To make a plan and try it out. (chapter 6) You can investigate to find out what snails eat.

make a model To make something to show how something looks or works. (chapter 5) You can make a model to show how a river flows into a lake.

measure To find out how far something moves, or how long, how much, or how warm something is. (chapter 9) You can use a thermometer to measure temperature.

observe To see, hear, taste, touch, or smell. (chapter 1) You can use your senses to observe a plant.

predict To use what you know to tell what you think will happen. (chapter 7) If you see storm clouds, you can predict that it will rain.

put things in order To tell or show what happens first, next, and last. (chapter 10) A chart can help you put things in order.

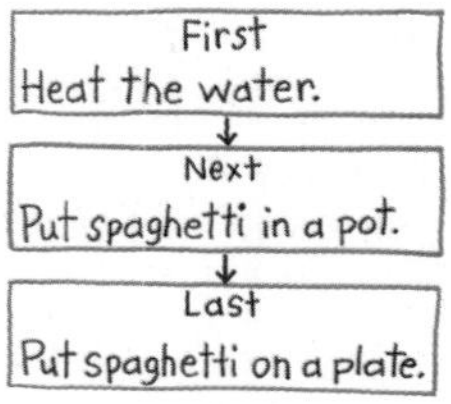

record data To write down what you observe. (chapter 8) You can record data about what snails ate.

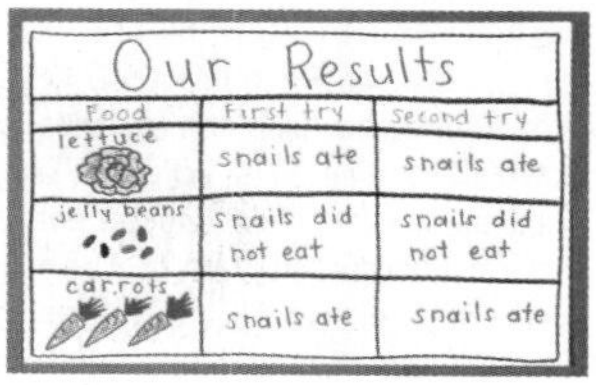

Credits

COV Juniors Bildarchiv/Alamy; **COV Back** (t)PhotoLink/Getty Images, (c)Nigel Cattlin/Holt Studios International Ltd/Alamy, (b)Iconotec/Glow Images; **COV Spine** Siede Preis/Getty Images; **i** Iconotec/Glow Images; **ii-iii** Photodisc/Getty Images; **iv** (tr)Dynamic Graphics Group/Creatas/Alamy, (tl)Ingram Publishing/Alamy, (b)Art Wolfe/The Image Bank/Getty Images; **vi** Tetsuya Tanooka/a.collectionRF/Getty Images; **vi-vii** Organics Image Library/Alamy; **vii** Design Pics RF/Getty Images; **viii** ©William Manning/Corbis; **x** (bl)Ken Cavanaugh/McGraw-Hill Education, (r)McGraw-Hill Education; **x-xi** Jules Frazier/Getty Images; **xi** David Buffington/Getty Images, (b)Patrick Orton/Aurora Open/Getty Images; **xii** (bl)Ted Clutter/Science Source, (t)C Squared Studios/Getty Images, (br)Dynamic Graphics Group/IT Stock Free/Alamy; **xiii** (t)Natalie Ray/McGraw-Hill Education, (b)Natalie Ray/McGraw-Hill Education; **Xiii** (tl)Natalie Ray/McGraw-Hill Education; **xiv** (t)Ken Cavanaugh/McGraw-Hill Education, (b)Mark Steinmetz/McGraw-Hill Education; **ix** (t)PhotoLink/Getty Images, (b)Blend Images/SuperStock, (inset)Creatas/SuperStock; **1** ©Little Blue Wolf Productions/Corbis; **2-3** Art Wolfe/The Image Bank/Getty Images; **3** (t to b)G.K. & Vikki Hart/Getty Images, (1)Photodisc/Getty Images, (2)G.K. & Vikki Hart/Getty Images, (3)Stockdisc/PunchStock, (4)Photodisc/Getty Images, (5)Photodisc/Getty Images; **4** (bl)Rosemary Calvert/Getty Images, (br)Creatas/PunchStock, (tl)Image Source/PunchStock, (tr)©Tobias Bernhard/Corbis; **5** (tr)Art Wolfe/The Image Bank/Getty Images, (bl)©Craig Tuttle/Corbis, (tl)John White Photos/Getty Images, (br)Suzanne L. and Joseph T. Collins/Science Source; **6** (c)G.K. & Vikki Hart/Getty Images, (b)Creatas/PunchStock, (cr)Photodisc/Getty Images, (cl)Brand X Pictures/PunchStock, (t)Stockbyte; **7** (t)Ken Cavanaugh/McGraw-Hill Education, (b)McGraw-Hill Education; **8** (l)©Gary Carter/Corbis, (r)Adam Jones/The Image Bank/Getty Images, (c)Rosemary Calvert/SuperStock; **9** (t)Natalie Ray/McGraw-Hill Education, (b)jaki good photography - celebrating the art of life/Moment Open/Getty Images; **10-11** ©Rob C. Nunnington/Gallo Images/Corbis; **11** (t)Trevor Smithers ARPS/Alamy, (b)Emmanuel Lattes/Alamy; **12** (b)Natalie Ray/McGraw-Hill Education, (t)Dynamic Graphics Group/IT Stock Free/Alamy; **13** Natalie Ray/McGraw-Hill Education; **14** (b)Natalie Ray/McGraw-Hill Education, (t)C Squared Studios/Getty Images; **15** (b)G.K. & Vikki Hart/Getty Images, (t)Natalie Ray/McGraw-Hill Education; **16** (tl)Michael Groen/McGraw-Hill Education, (br)Ken Cavanaugh/McGraw-Hill Education, (tr)Ken Cavanaugh/McGraw-Hill Education, (c)Ken Cavanaugh/McGraw-Hill Education, (bl)Ken Cavanaugh/McGraw-Hill Education; **17** Cephas Picture Library/Alamy; **18** Steve Hopkin/The Image Bank/Getty Images; **19** (t)Dan Suzio/Science Source, (b)Simon Clairac/Alamy; **20-21** Inspirational Images by Ken Hornbrook/Moment/Getty Images; **21** (tc)C. Borland/PhotoLink/Getty Images, (t)Comstock/PictureQuest, (bc)Evan Sklar/The Image Bank/Getty Images, (b)Maximilian Stock Ltd/Science Source; **22-23** Photo Japan/Alamy; **23** (t to b) McGraw-Hill Education, (1)McGraw-Hill Education, (2)Michael Scott/McGraw-Hill Education, (3)McGraw-Hill Education, (4)Natalie Ray/McGraw-Hill Education; **24-25** ©Glowimages/Corbis; **25** Ken Karp/McGraw-Hill Education; **26** (r)Digital Vision/PunchStock, (l)Wesley Hitt/Getty Images; **27** Iconotec/Glow Images; **28-29** Lorelei Mann/McGraw-Hill Education; **29** (b) McGraw-Hill Education, (t)McGraw-Hill Education, (c)Ken Cavanaugh/McGraw-Hill Education; **30** Brand X Pictures/PunchStock; **30-31** Siede Preis/Getty Images; **32** (t)Linda Holt Ayriss/Getty Images, (c)Maximilian Stock Ltd/Science Source, (b)Siede Preis/Getty Images; **34** Chris Cheadle/Alamy; **35** (t)Ken Cavanaugh/McGraw-Hill Education, McGraw-Hill Education, Ken Cavanaugh/McGraw-Hill Education, Ken Cavanaugh/McGraw-Hill Education, (tc)Ken Cavanaugh/McGraw-Hill Education, (b)Ken Cavanaugh/McGraw-Hill Education, Ken Cavanaugh/McGraw-Hill Education, (bc)McGraw-Hill Education; **36-37** Emilio Ereza/age fotostock; **37** (b)Natalie Ray/McGraw-Hill Education, (t)Ken Cavanaugh/McGraw-Hill Education, (c)McGraw-Hill Education; **38** (br)C Squared Studios/Getty Images, (t)Tetsuya Tanooka/a.collectionRF/Getty Images, (bl)5second/iStock/Getty Images Plus/Getty Images; **39** (t)©Harry Spurling/Royalty-Free/Corbis, (b)Organics Image Library/Alamy; **40** (bl)Paul Franklin/Oxford Scientific/Getty Images, (b)C Squared Studios/Getty Images, (br)©Nik Wheeler/Corbis, (t)C Squared Studios/Getty Images, (c)foodfolio/Alamy; **41** (b)McGraw-Hill Education, (t)Burke Triolo Productions/Getty Images; **42** (b)Spencer Grant/PhotoEdit, (t) Stockdisc/PunchStock; **42-43** ©Craig Lovell/Corbis; **43** Stockdisc/PunchStock; **44** (r)Stockdisc/PunchStock, (l)Claudia Uribe/Digital Vision/Getty Images; **45** (l)©Michael T.Sedam/Royalty-Free/Corbis, (r)McGraw-Hill Education; **46** (l)Evan Sklar/The Image Bank/Getty Images, (r)©Photodisc/Getty Images; **47** Ken Cavanaugh/McGraw-Hill Education, (t)Macmillan/McGraw-Hill Education, (b)Macmillan/McGraw-Hill Education, (c)Macmillan/McGraw-Hill Education; **48** (tl)Siede Preis/Getty Images, (cr)©Laura Sivell/Papilio/Corbis, (cl)Pixtal/age fotostock, (br)Maximilian Stock Ltd/Science Source, (bl)McGraw-Hill Education, (tr)Burke Triolo Productions/Getty Images; **49** (bl)©Carl & Ann Purcell/Corbis, (tr)Comstock/PictureQuest, (tl)C. Borland/PhotoLink/Getty Images, (br)©Noam Armonn/age fotostock; **50-51** Andy Roberts/OJO Images/Getty Images; **51** (tc)Bogdan Wankowicz/Alamy, (t)©Image Source/Corbis, (b)Perytskyy/iStock/Getty Images Plus/Getty Images, (c)tonda/iStock/Getty Images Plus/Getty Images; **52-53** Bartomeu Borrell/age fotostock; **53** (b)Natalie Ray/McGraw-Hill Education, (bc)Joe Polillio/McGraw-Hill Education, (t)McGraw-Hill Education, (tc)Ken Cavanaugh/McGraw-Hill Education; **54** ©Jose Fuste Raga/Corbis; **55** (tl)©Image Source/Corbis, (tr)Gilbert S. Grant/Science Source, (b)Judd Pilossof/Photolibrary/Getty Images; **56** Michael Scott/McGraw-Hill Education; **57** (t)©Niall Benvie/Corbis, (b)Paul McCormick/The Image Bank/Getty Images; **58-59** Steve Satushek/The Image Bank/Getty Images; **59** (t to b)McGraw-Hill Education, (1)Ken Cavanaugh/McGraw-Hill Education, (2)McGraw-Hill Education, (3)Jacques Cornell/McGraw-Hill Education, (4)Ken Karp/McGraw-Hill Education; **60** (b)Nigel Cattlin/Alamy, (t)Anne Hyde/Getty Images, (bc)Nigel Cattlin/Alamy, (r)Nigel Cattlin/Alamy; **60-61** Siede Preis/Getty Images; **61** (b)Bogdan Wankowicz/Alamy, (t)WizData, Inc./Alamy; **62** (b)Natalie Ray/McGraw-Hill Education, (t) Natalie Ray/McGraw-Hill Education; **63** Natalie Ray/McGraw-Hill Education; **64** (t)Annie Reynolds/PhotoLink/Getty Images, (br)Melanie Acevedo/The Image Bank/Getty Images, (bl)NSP-RF/Alamy; **65** SuperStock; **66-67** Claude Dagenais/iStock/Getty Images Plus/Getty Images; **67** (b)Ken Karp/McGraw-Hill Education, (t)Ken Karp/McGraw-Hill Education, (c)McGraw-Hill Education; **68** (t)tonda/iStock/Getty Images Plus/Getty Images; **69** (b)Nolmedrano99/iStock/Getty Images Plus/Getty Images, (inset)Nolmedrano99/iStock/Getty Images Plus/Getty Images; **70** PhotoLink/Getty Images, (c)Juniors Bildarchiv/Alamy; **70-71** Blake Kent/Design Pics/Getty Images; **71** Perytskyy/iStock/Getty Images Plus/Getty Images; **72** ©Michael Boys/Corbis; **73** (l)C Squared Studios/Getty Images, (c)C Squared Studios/Getty Images, (r)Ingram Publishing/SuperStock; **74** age fotostock/SuperStock; **75** ©Arvind Garg/Corbis; **76** Nick White/Digital Vision/Getty Images; **77** Dagny Willis/Moment/Getty Images; **78** (l to r)Nigel Cattlin/Alamy, (1)Bogdan Wankowicz/Alamy, (2)Nigel Cattlin/Alamy, (3)Nigel Cattlin/Alamy, (4)Dynamic Graphics/Punchstock, (5)D. Hurst/Alamy, (6)©Royalty-Free/Corbis, (7)travis manley/iStock/Getty Images Plus/Getty Images, (bkgd)Siede Preis/Getty Images; **79** (l to r)Stockdisc/Punchstock, (1)Purestock/SuperStock, (2)Stockbyte/PunchStock, (3)Photolibrary/Alamy, (4)©Maximilian Stock Ltd/PhotoCuisine/Corbis, (5)PhotoSpin, Inc/Alamy, (6)lynx/iconotec.com/Glow Images, (7)©Goodshoot/Corbis; **80** (bl)©Holger Winkler/Corbis, (t)John Lund/Sam Diephuis/Blend Images/Getty Images Plus/Getty Images, (br)©Jonathan Blair/Corbis; **81** Bob Elsdale/The Image Bank/Getty Images; **84** Jacques Cornell/Macmillan/McGraw-Hill; **84-85** Dave Cole/Alamy; **85** (bc)IT Stock/Punchstock, (t)©DLILLC/Corbis, (tc)Johner Images/Alamy, (b)©MedioImages/SuperStock; **86-87** ©Dale Spartas/Corbis; **87** (l to r)James Urbach/SuperStock, (1)Image Source/Punchstock, (2)Brand X Pictures/PunchStock, (3)D. Hurst / Alamy, (4)Ken Cavanaugh/McGraw-Hill Education, (5)Jacques Cornell/McGraw-Hill Education; **88** (l)Robert Muckley/Getty Images, (r)©DLILLC/Corbis; **89** (b)Photodisc/Getty Images, (t)Johner Images/Alamy, (c)by_nicholas/iStock/Getty Images Plus/Getty Images; **90** (t)IT Stock/Punchstock, (c)Jack Goldfarb/Design Pics; **90-91** Photodisc/Getty Images; **91** (r)©MedioImages/SuperStock, (l)©Ingram Publishing/Alamy; **92** (c)Stockdisc(Stockbyte)/Getty Images, (b)Photodisc Collection/Getty Images, (inset)mychadre77/iStock/Getty Images Plus/Getty Images, (t)Imagestate/Alamy; **93** (c)Ted Clutter/Science Source, (b)Burke/Triolo Productions/Brand X Pictures/Getty Images, (t)Photodisc/Getty Images, (inset)Dynamic Graphics Group/IT Stock Free/Alamy; **94-95** NPS Photo by Willis Peterson; **95** (t to b)McGraw-Hill Education, (1)McGraw-Hill Education, (2)McGraw-Hill Education, (3)Photodisc/Getty Images, (4)Ken Cavanaugh/McGraw-Hill Education, (5)Natalie Ray/McGraw-Hill Education; **96** Art Wolfe/The Image Bank/Getty Images; **97** (t)Sean Russell/Getty Images, (b)Westend61/Getty Images; **98** (b)Norbert Rosing/National Geographic/Getty Images, (t)Theo Allofs/Photodisc/Getty Images; **99** Martin Ruegner/Radius Images/Getty Images; **100** (r)Ken Karp/McGraw-Hill Education, (l)G.K. & Vikki Hart/Getty Images; **102-103** Jane Shauck Photography/Alamy; **103** (t to b)Ken Cavanaugh/McGraw-Hill Education, (1)Ken Cavanaugh/McGraw-Hill Education, (2)McGraw-Hill Education, (3)McGraw-Hill Education, (4)Ken Karp/McGraw-Hill Education; **104-105** Ingram Publishing; **105** Diane Diederich/iStock/Getty Images Plus/Getty Images; **106** Juniors Bildarchiv GmbH/Alamy; **107** Karen Doody/Stocktrek Images/Getty Images; **108-109** Martin Rugner/age fotostock; **109** (l to r)Ingram Publishing/Alamy, (1)G.K. & Vikki Hart/Getty Images, (2)©Peter M. Fisher/Corbis, (3)Photodisc/Getty Images, (4)Ingram Publishing/Alamy, (5)Stockbyte/Punchstock; **110** (l)Gerard Brown/Dorling Kindersley/Getty Images, (r)Les Stocker/Oxford Scientific/Getty Images; **111** (l)Les Stocker/Oxford Scientific/Getty Images, (r)Fuse/Getty Images; **112** (l)Dorling Kindersley/Getty Images, (r)©kerkla /iStockphoto.com/Getty Images; **113** (l)Herman du Plessis/Gallo Images/Getty Images, (r)bazilfoto/iStock/Getty Images Plus/Getty Images; **114** (cl)David Boag/Oxford Scientific/Getty Images, (cr)Papilio/Alamy, (l)bulentozber/iStock/Getty Images Plus/Getty Images, (r)Papilio/Alamy; **114-115** (t)Oleg Moiseyenko/Alamy, (b)©ASO FUJITA/amanaimagesRF/amanaimages/Corbis; **115** Digital Vision/Punchstock; **116** Courtesy of American Museum of Natural History; **116-117** Andre_BR/iStock/Getty Images Plus/Getty Image; **117** Natural Visions/Alamy; **118** (t)Exactostock/SuperStock, (b)Chatterer/Getty Images; **119** (b)Blickwinkel/Alamy, (t)Sam Lee/Getty Images; **120** (t)Stockbyte/Getty Images, (b)Yuri Arcurs/Design Pics; **121** (b)©Nigel J. Dennis/Gallo Images/Corbis, (t)©Juniors Bildarchiv/Alamy; **122** (tr)IT Stock/Punchstock, (bl)©MedioImages/SuperStock, (cr)Johner Images/Alamy, (cl)Stockdisc(Stockbyte)/Getty Images, (br)Robert Muckley/Getty Images, (tl)Dynamic Graphics Group/IT Stock Free/Alamy; **123** (r)Dynamic Graphics Group/Creatas/Alamy, (l)Theo Allofs/Photodisc/Getty Images; **124-125** (t)Sam Lee/Getty Images; **125** (cr)©Arthur Morris/Corbis, (tc)Spiraling Road Photography/GettyImages, (bcl)David R. Frazier Photolibrary, Inc./Alamy, (inset)John Warden/The Image Bank/Getty Images, (b)Jeff Hunter/Getty Images, (t)©iStockphoto.com/Andyworks; **126-127** ©W.Perry Conway/Corbis; **127** (t to b)Ken Cavanaugh/McGraw-Hill Education, (1)McGraw-Hill Education, (2)Jacques Cornell/McGraw-Hill Education, (3)Joe Polillio/McGraw-Hill Education, (4)Natalie Ray/McGraw-Hill Education; **128** ©iStockphoto.com/Andyworks; **128-129** Nigel J. Dennis/Science Source; **130** (br)Getty Images, (t)Natalie Ray/McGraw-Hill Education, (bl)Spiraling Road Photography/GettyImages; **131** Nicole Duplaix/National Geographic/Getty Images; **132** Burke/Triolo Productions/Brand X Pictures/Getty Images; **132-133** Jim Zipp/Science Source; **133** (t to b)Jacques Cornell/McGraw-Hill Education, (1)Jacques Cornell/McGraw-Hill Education, (2)McGraw-Hill Education, (3)Michael Scott/McGraw-Hill Education, (4)McGraw-Hill Education, (5)Natalie Ray/McGraw-Hill Education; **134** John Warden/The Image Bank/Getty Images; **134-135** David R. Frazier Photolibrary, Inc./Alamy; **135** (r)Harry Engels/Science Source, (l)©Arthur Morris/Corbis, (b)R Sherwood Veith/Getty Images; **136** (t)Georgette Douwma/Science Source, (b)Jeff Hunter/Getty Images; **137** Douglas Klug/Getty Images; **138** Comstock Images/Stockbyte/Getty Images; **139** (bl)©Punchstock/Corbis, (tr)Andrey Nekrasov/imageBROKER/Getty Images, (br)bookguy/iStock/Getty Images Plus/Getty Images, (tl)Cigdem Sean Cooper/Moment/Getty Images; **140-141** Brad Boland/Flickr Flash/Getty Images; **141** (b)Comstock Images/Alamy, (t)Ken Cavanaugh/McGraw-Hill Education, (c)McGraw-Hill Education; **142** (b)Design Pics RF/Getty Images, (t)iStockphoto/Getty Images; **143** (t)Imagebroker/Alamy, (b)©Lothar Lenz/Corbis; **144** (tr)Digital Vision/Getty Images, (l)ARCO/Reinhard H/age fotostock, (c)©Ingram Publishing/age fotostock; **144-145** Frank Greenaway/DK Images, ©Fritz Rauschenbach/Corbis; **145** Nigel McCall/Alamy; **146** (l)Digital Vision/PunchStock, (r)Steve Maslowski/Science Source; **148** (b)American Museum of Natural History, (t)Courtesy of Jin Meng; **149** Craig Chesek/American Museum of Natural History; **150** (bkgd)Digital Vision/Getty Images, (inset)Getty Images; **151** Art Wolfe/The Image Bank/Getty Images; **152** age fotostock/SuperStock; **153** (t)ImageState/Alamy, (b)Gary Yeowell/The Image Bank/Getty Images; **154** (br)William Ervin/Science Source, (t)©Fritz Rauschenbach/Corbis, (tr)Nigel McCall/Alamy, (tcr)©Ingram Publishing/age fotostock, (tl)ARCO/Reinhard H/age fotostock; **155** (tr)©Arthur Morris/Corbis, (c)John Warden/The Image Bank/Getty Images, (b)Digital Vision/PunchStock, (tl)James Randklev/Getty Images, (bcl)David R. Frazier Photolibrary, Inc./Alamy; **156** (t)Thinkstock/Alamy, (br)©Bernd Obermann/Corbis, (bl)Yoel Harel/Alamy; **157** ©Craig Lovell/Corbis; **160-161** Dynamic Graphics Group/Creatas/Alamy; **161** (t)A & C Wiley/Photolibrary/Getty Images, (b)©Don Hammond/age fotostock, (bc)Gary Yeowell/The Image Bank/Getty Images, (tc)A & C Wiley/Photolibrary/Getty Images; **162-163** Takumi Terao/Moment Open/Getty Images; **163** (b)Natalie Ray/McGraw-Hill Education, (c)McGraw-Hill Education, (t)McGraw-Hill Education; **164** Index Stock Imagery, Inc; **165** NASA/age fotostock; **166** (b)©Larry Dale Gordon/Corbis, (t)Ken Cavanaugh/McGraw-Hill Education; **167** (r)A & C Wiley/Photolibrary/Getty Images, (l)Radius Images/Radius Images/Getty Images Plus/Getty Images; **168** Gary Yeowell/The Image Bank/Getty Images; **168-169** Don Hammond/age fotostock; **170-171** Joyce Photographics/Science Source; **171** (b)Natalie Ray/McGraw-Hill Education, (c)Ken Cavanaugh/McGraw-Hill Education, (t)McGraw-Hill Education; **172** (tr)Ken Cavanaugh/McGraw-Hill Education, (tl)Ken Cavanaugh/McGraw-Hill Education, (br)Ken Cavanaugh/McGraw-Hill Education, (bl)Jacques Cornell/McGraw-Hill Education; **173** (c)Tony Lilley/Alamy, (tr)Siede Preis/Getty Images, (tl)Siede Preis/Getty Images, (cr)Jacques Cornell/McGraw-Hill Education, (cl)Jacques Cornell/McGraw-Hill Education, (b)Jacques Cornell/McGraw-Hill Education; **174** (tr)Natalie Ray/McGraw-Hill Education, (cr)Natalie Ray/McGraw-Hill Education, (br)Natalie Ray/McGraw-Hill Education, (tl)Natalie Ray/McGraw-Hill Education, (inset)Natalie Ray/McGraw-Hill Education, (bl)Natalie Ray/McGraw-Hill Education, (cl)Natalie Ray/McGraw-Hill Education; **175** RNMitra/Getty Images; **176** Denis Finnin/American Museum of Natural History; **176-177** Martin Rogers/The Image Bank/Getty Images; **177** (b)PhotoLink/Getty Images, (t)©Matthias Kulka/Corbis; **178-179** ©David Samuel Robbins/Corbis; **179** (b)Natalie Ray/McGraw-Hill Education, (c)Michael Scott/McGraw-Hill Education, (t)McGraw-Hill Education; **180** Wilfried Krecichwost/Stockbyte/Getty Images; **181** Mark Lewis/Photographer's Choice/Getty Images; **182** ©Pixtal/age fotostock; **182-183** ©William Manning/Corbis; **183** (b)©William Manning/Corbis, (t)C Squared Studios/Getty Images; **184** Sotirios Milionis/Alamy; **185** (l)Ken Cavanaugh/McGraw-Hill Education, (r)Brand X Pictures/PunchStock; **186-187** Medioimages/PunchStock; **187** Creatas/Punchstock; **188** Duncan Maxwell/Robert Harding World Imagery/Getty Images; **189** (t)©Hubert Stadler/Corbis, (b)Corbis/PunchStock; **190** (c)H. Wiesenhofer/Photolink/Getty Images, (t)A & C Wiley/Photolibrary/Getty Images, (b)John Elk III/Lonely Planet Images/Getty Images; **191** (l)Jacques Cornell/McGraw-Hill Education, (c)Jacques Cornell/McGraw-Hill Education, (r)Ken Cavanaugh/McGraw-Hill Education; **192-193** Rachel Weill/UpperCut Images/Getty Images; **193** (t)Deco/Alamy, (bc)BigCheesePhoto/PunchStock, (b)Photka/iStock/Getty Images Plus/Getty Images, (tc)Natalie Ray/McGraw-Hill Education; **194-195** Natalie Ray/McGraw-Hill Education; **195** (b)Natalie Ray/McGraw-Hill Education, (t)McGraw-Hill Education; **196** Ron Rovtar Photography/Alamy; **196-197** ©Rainer Hackenberg/Corbis; **197** (r)©Rose Hartman/Corbis, (l)John James Wood/Photolibrary/Getty Images; **198** (b)George and Monserrate Schwartz/Alamy, (t)Photodisc/Getty Images; **199** (t)Mel Curtis/Getty Images, (b)Designer491/iStock/Getty Images Plus/Getty Images; **200-201** ©Bettmann/Corbis; **201** (b)McGraw-Hill Education, (t)Ken Cavanaugh/McGraw-Hill Education, (c)McGraw-Hill Education; **202** (r)Comstock/Punchstock, (l)Andrew Hill/iStock/Getty Images Plus/Getty Images; **203** (l)Peter Cade/The Image Bank/Getty Images, (r)GeorgeSe/iStock/Getty Images Plus/Getty Images; **204** Deco/Alamy; **205** ©Dean Conger/Corbis; **206** Courtesy of Mark Siddall; **207** (tl)Denis Finnin/American Museum of Natural History, (tr)Courtesy of Mark Siddall/American Museum of Natural History, (b)Colin Keates/Dorling Kindersley/Getty Images; **208-209** David Seed Photography/The Image Bank/Getty Images; **209** (b)Natalie Ray/McGraw-Hill Education, (bc)Jacques Cornell/McGraw-Hill Education, (t)McGraw-Hill Education, (tc)McGraw-Hill Education; **210** Natalie Ray/McGraw-Hill Education; **211** (b)Natalie Ray/McGraw-Hill Education, (t)Ken Cavanaugh/McGraw-Hill Education; **212** (r)BigCheesePhoto/PunchStock, (l)Natalie Ray/McGraw-Hill Education; **213** Photka/iStock/Getty Images Plus/Getty Images; **214** Steffen Thalemann/Stockbyte/Getty Images; **215** C Squared Studios/Getty Images; **216** Joanna Vidad/Hemera/Getty Images Plus/Getty Images; **217** Reed Kaestner/Getty Images; **218** ©Nik Wheeler/Corbis; **219** Natalie Ray/McGraw-Hill Education; **220** (r)Steve Allen/Brand X Pictures/Alamy, (l)©Dean Conger/Corbis; **221** (t)Nigel Cattlin/Science Source, (b)Photka/iStock/Getty Images Plus/Getty Images; **222** (t)©Stephen Frink/Corbis, (bl)Henry Chaplin/iStock/Getty Images Plus/Getty Images, (br)©Paul A. Souders/Corbis; **223** IMAGINA Photography/Alamy; **224** ramzihachicho/Getty Images; **225** (c)©PictureQuest/Corbis, (b)Ron Niebrugge/Alamy, (t)©Royalty-Free/Corbis; **226-227** ©Royalty-Free/Corbis; **227** (bc)Ariel Skelley/Getty Images, (tc)Photodisc/PunchStock, (b)Brand X Pictures/PunchStock, (t)©Richard Cummins/Corbis; **228-229** Adam Burn/Getty Images; **229** (b)Natalie Ray/McGraw-Hill Education, (tc)Ken Cavanaugh/McGraw-Hill Education, (t)Ken Cavanaugh/McGraw-Hill Education, (bc)Ken Cavanaugh/McGraw-Hill Education; **230** (l)©G. Schuster/Corbis, (r)©Royalty-Free/Corbis; **231** (bl)Robert Brenner/PhotoEdit, (t)Brand X Pictures/PunchStock, (br)Rob Casey/The Image Bank/Getty Images; **232** (t to b)Don Farrall/Getty Images, (1)Jacques Cornell/McGraw-Hill Education, (2)Pixtal/age fotostock, (3)PhotoLink/Getty Images, (4)Creatas/SuperStock; **233** Ed-Imaging/McGraw-Hill Education; **234-235** ©Daryl Benson/Masterfile/Corbis; **235** (b)McGraw-Hill Education, (t)McGraw-Hill Education, (c)Ken Cavanaugh/McGraw-Hill Education; **237** (bl)Altrendo Images/Altrendo/Getty Images, (br)Scott Sroka/National Geographic/Getty Images, (t)Uli Nusko/Alamy; **238** (b)©Craig Aurness/Corbis, (t)McGraw-Hill Education, (c)©Punchstock/Corbis; **239** Jason Weingart Photography; **240-241** Connie Coleman/Photographer's Choice/Getty Images; **241** (t to b)McGraw-Hill Education, (1)McGraw-Hill Education, (2)McGraw-Hill Education, (3)Ken Cavanaugh/McGraw-Hill Education, (4)Natalie Ray/McGraw-Hill Education; **242** Caiaimage/Paul Bradbury/OJO+/Getty Images; **243** (b)©Richard Cummins/Corbis, (t)Brand X Pictures/PunchStock; **244** Photodisc/PunchStock; **245** Paul Gapper/Alamy; **246** David Sanger Photography/Alamy; **246-247** age fotostock/SuperStock; **248-249** Ariel Skelley/Getty Images; **249** (t to b)McGraw-Hill Education, (1)McGraw-Hill Education, (2)McGraw-Hill Education, (3)Jacques Cornell/McGraw-Hill Education, (4)Natalie Ray/McGraw-Hill Education; **250** Alida Thorpe/Moment/Getty Images; **251** (bl)©Chase Swift/Corbis, (t)C Squared Studios/Getty Images, (br)Westend61 GmbH/Alamy; **252** (r)Ray Sandusky/Brentwood, TN/Moment/Getty Images, (l)Westend61 - FotoFealing/Brand X Pictures/Getty Images, (t)Aluma Images/Stockbyte/Getty Images, (b)Elenathewise/Getty Images; **253** Owen Newman/Oxford Scientific/Getty Images; **254** (l)Brand X Pictures/PunchStock, (r)Brand X Pictures/PunchStock; **255** Jules Frazier/Getty Images; **256** Blend Images/SuperStock; **257** Mike Powell/The Image Bank/Getty Images; **258** ©Ingram Publishing/SuperStock; **259** Richard Hutchings/PhotoEdit; **260** (b)McGraw-Hill Education, (bc)Rob Casey/The Image Bank/Getty Images, (tc)©Punchstock/Corbis, (t)©G. Schuster/Corbis; **261** (tl)Ron Niebrugge/Alamy, (tr)John E. Marriot/Alamy, (b)Jack Sullivan/Alamy; **262-263** Karl Lehmann/Lonely Planet Images/Getty Images; **263** (bc)Eckhard Slawik/Science Source, (t)Marius Hepp/EyeEm/EyeEm/Getty Images, (tc)©G. Schuster/Corbis; **264-265** Dragan Romanovic/iStock/Getty Images Plus/Getty Images; **265** Blend Images/Alamy; **266-267** Chris Stein/Photodisc/Getty Images, (bkgd)Marius Hepp/EyeEm/EyeEm/Getty Images; **267** (b)Jerry Lodriguss/Science Source, (t)C Squared Studios/Getty Images; **268** (l)Comstock/PunchStock, (r)PCL/Alamy; **269** Images Etc Ltd/The Image Bank/Getty Images; **270-271** ohnnya123/iStock/Getty Images Plus/Getty Images; **271** (t to b)Ken Cavanaugh/McGraw-Hill Education, (1)Jacques Cornell/McGraw-Hill Education,

(2)McGraw-Hill Education, (3)Ken Cavanaugh/McGraw-Hill Education, (4)McGraw-Hill Education, (5)Natalie Ray/McGraw-Hill Education; **272** (l)IT Stock Free/Punchstock, (r)Leander Baerenz/The Image Bank/Getty Images; **275** (l)Natalie Ray/McGraw-Hill Education, (cl)Natalie Ray/McGraw-Hill Education, (r)Natalie Ray/McGraw-Hill Education, (cr)Natalie Ray/McGraw-Hill Education; **276** PhotoLink/Getty Images; **277** Brand X Pictures/PunchStock; **278-279** Stockbyte/PunchStock; **279** (l to r)Eckhard Slawik/Science Source, (1)Eckhard Slawik/Science Source, (2)Eckhard Slawik/Science Source, (3)Eckhard Slawik/Science Source; **280** Richard Wahlstrom/ The Image Bank/Getty Images; **281** (inset)McGraw-Hill Education, (cl)Eckhard Slawik/Science Source, (c)NASA/age fotostock, (bkgd)Shaun Lowe/Getty Images, (b)Eckhard Slawik/Science Source, (t)Eckhard Slawik/Science Source, (cr)Eckhard Slawik/Science Source; **284** Denis Finnin/American Museum of Natural History; **284-285** (bkgd)Ido Greiman/Getty Images; **285** NASA/JPL-Caltech; **286-287** Guillaume Seguin/Moment/Getty Images; **288** PhotoLink/Getty Images; **288-289** ©Royalty-Free/Corbis; **291** (tl)Andre Gallant/The Image Bank/Getty Images, (bl)Eckhard Slawik/Science Source, (bc)Eckhard Slawik/Science Source, (br)Eckhard Slawik/Science Source, (tr)Liesel Bockl/fStop/Getty Images; **292** (bl)Digital Vision/Getty Images, (br)Dennis MacDonald/Alamy, (t)©Mike Hollingshead/Corbis; **293** Emery,Steven/Photolibrary/Getty Images; **294** ©Jonathan Andrew/Corbis; **295** (c)Emma Lee/Life File/Getty Images, (b)©Klaus Hackenberg/Corbis, (t)Anthony Chatfield/EyeEm/Getty Images; **296-297** Frans Lemmens/The Image Bank/Getty Images; **297** (t)Jacques Cornell/McGraw-Hill Education, (tc)Ken Karp/McGraw-Hill Education, (b)Jules Frazier/Getty Images, (bc)Jill Braaten/McGraw-Hill Education; **298-299** Ken Cavanaugh/McGraw-Hill Education; **299** (b)Natalie Ray/McGraw-Hill Education, (t)McGraw-Hill Education, (bc)McGraw-Hill Education, (tc)McGraw-Hill Education, (bkgd)Jules Frazier/Getty Images; **300** (l)©Ingram Publishing/Alamy, (r)D. Hurst/Alamy; **301** (tl)Photodisc/Getty Images, (b)Buzzshotz/Alamy, (tr)Ken Cavanaugh/McGraw-Hill Education; **302** Natalie Ray/McGraw-Hill Education; **303** Natalie Ray/McGraw-Hill Education; **304** Ken Cavanaugh/McGraw-Hill Education; **305** (l)Natalie Ray/McGraw-Hill Education, (c)Natalie Ray/McGraw-Hill Education, (r)Natalie Ray/McGraw-Hill Education; **306-307** Natalie Ray/McGraw-Hill Education; **307** (b)Natalie Ray/McGraw-Hill Education, (t)McGraw-Hill Education, (c)Jacques Cornell/McGraw-Hill Education; **308** Ken Karp/McGraw-Hill Education; **309** (b)Ken Karp/McGraw-Hill Education, (t)Natalie Ray/McGraw-Hill Education; **310** (t)McGraw-Hill Education, (b)Ken Cavanaugh/McGraw-Hill Education; **311** (t)McGraw-Hill Education, (bc)McGraw-Hill Education, (tc)Burke/Triolo/Brand X Pictures/Jupiterimages, (b)McGraw-Hill Education; **312** (b)Glow Images, (t)Image Source/PunchStock, (c)apomares/iStock/Getty Images Plus/Getty Images; **313** (b)Chloe Johnson/Alamy, (tc)Steve Gorton/Dorling Kindersley/Getty Images, (bc)Ingram Publishing, (t)Darren Greenwood/Design Pics; **314-315** Steve Murez/Photographer's Choice/Getty Images; **315** (t to b)Jacques Cornell/McGraw-Hill Education, (1)Ken Karp/McGraw-Hill Education, (2)Jacques Cornell/McGraw-Hill Education, (3)Jacques Cornell/McGraw-Hill Education, (4)Ken Karp/McGraw-Hill Education; **316** (tl)Groesbeck/Uhl/Getty Images, (tr)Masterfile (Royalty-Free Div.), (bl)Jill Braaten/McGraw-Hill Education, (br)Martyn F. Chillmaid/Science Source; **316-317** Natalie Ray/McGraw-Hill Education; **317** (c)Ken Cavanaugh/McGraw-Hill Education, (b)Ken Cavanaugh/McGraw-Hill Education, (t)Ken Cavanaugh/McGraw-Hill Education; **319** Joseph McNally/Photodisc/Getty Images; **320-321** Andre Jenny/Alamy; **322** Martyn Vickery/Alamy; **323** Leah Warkentin/Design Pics; **324** (tr)Ken Karp/McGraw-Hill Education, (bl)Jacques Cornell/McGraw-Hill Education, (tl)Martyn F. Chillmaid/Science Source, (br)Jules Frazier/Getty Images; **325** (bc)Groesbeck/Uhl/Getty Images, (bl)Masterfile (Royalty-Free Div.), (br)Jill Braaten/McGraw-Hill Education; **326-327** Leah Warkentin /Design Pics; **327** (t)Dennis Gray/Cole Group/Getty Images, (b)McGraw-Hill Education, (tc)Natalie Ray/McGraw-Hill Education; **328-329** Natalie Ray/McGraw-Hill Education; **329** (b)Natalie Ray/McGraw-Hill Education, (bc)McGraw-Hill Education, (tc)Ken Cavanaugh/McGraw-Hill Education, (t)Jacques Cornell/McGraw-Hill Education; **330** Natalie Ray/McGraw-Hill Education; **331** (t)Ken Cavanaugh/McGraw-Hill Education, (bl)Ken Cavanaugh/McGraw-Hill Education, (br)Ken Karp/McGraw-Hill Education; **332-333** incamerastock/Alamy; **333** (t to b)Ken Cavanaugh/McGraw-Hill Education, (1)Ken Cavanaugh/McGraw-Hill Education, (2)Ken Cavanaugh/McGraw-Hill Education, (3)McGraw-Hill Education, (4)McGraw-Hill Education, (5)Natalie Ray/McGraw-Hill Education; **334** Natalie Ray/McGraw-Hill Education; **335** (bl)Natalie Ray/McGraw-Hill Education, (br)Natalie Ray/McGraw-Hill Education, (t)C Squared Studios/Getty Images; **336** (l)Natalie Ray/McGraw-Hill Education, (r)Natalie Ray/McGraw-Hill Education; **337** Natalie Ray/McGraw-Hill Education; **338** Ken Cavanaugh/McGraw-Hill Education; **339** foodfolio/Alamy; **340-341** Wides & Holl/The Image Bank/Getty Images; **341** (b)Natalie Ray/McGraw-Hill Education, (t)McGraw-Hill Education, (tc)Jacques Cornell/McGraw-Hill Education, (bc)Ken Cavanaugh/McGraw-Hill Education; **342** (l)Ken Karp/McGraw-Hill Education, (c)Ken Karp/McGraw-Hill Education, (r)Ken Karp/McGraw-Hill Education; **343** (bl)Ken Karp/McGraw-Hill Education, (bc)McGraw-Hill Education, (br)Ken Karp/McGraw-Hill Education, (t)Jacques Cornell/McGraw-Hill Education; **344** (l)©Luis Forra/epa/Corbis, (r)©Luis Forra/epa/Corbis; **345** Natalie Ray/McGraw-Hill Education; **346** Hemera Technologies/PhotoObjects.net/Getty Images Plus/Getty Images; **346-347** Photodisc/PunchStock; **347** lynx/iconotec.com/Glow Images; **348-349** C Squared Studios/Getty Images; **350** (t)Dennis Gray/Cole Group/Getty Images, (b)©Ben Fink/Brand X/Corbis; **351** (t)Mitch Hrdlicka/Getty Images, (b)©Royalty-Free/Corbis; **352** (bc)Ken Karp/McGraw-Hill Education, (b)Ken Karp/McGraw-Hill Education, (t)Natalie Ray/McGraw-Hill Education, (tc)Natalie Ray/McGraw-Hill Education; **353** (t)Christina Kennedy/PhotoEdit, (cl)Ken Cavanaugh/McGraw-Hill Education, (cr)Ken Cavanaugh/McGraw-Hill Education, (b)Natalie Ray/McGraw-Hill Education; **354** (t)Rosenfeld Images Ltd/Science Source, (br)Justin Kase/Alamy, (bl)RubberBall/SuperStock; **355** Purple Marbles/Alamy; **358-359** Christina Kennedy/PhotoEdit; **359** (tc)Michael Scott/McGraw-Hill Education, (bc)Photofusion Picture Library/Alamy, (b)Ken Cavanaugh/McGraw-Hill Education, (t)Chris Clinton/The Image Bank/Getty Images; **360-361** Robert Warren/Taxi/Getty Images; **361** (b)Natalie Ray/McGraw-Hill Education, (t)McGraw-Hill Education; **364** (bl)Brand X Pictures/Punchstock, (br)Thinkstock/Punchstock, (t)Ken Karp/McGraw-Hill Education; **365** Digital Vision/Getty Images; **366-367** ©Joyce Choo/Corbis; **367** (b)Natalie Ray/McGraw-Hill Education, (c)McGraw-Hill Education, (t)Janette Beckman/McGraw-Hill Education; **368** (r)Michael Scott/McGraw-Hill Education, (l)Chris Clinton/The Image Bank/Getty Images; **369** (bl)Comstock Images/Punchstock, (t)Natalie Ray/McGraw-Hill Education, (br)Natalie Ray/McGraw-Hill Education, (bkgd)Natalie Ray/McGraw-Hill Education, (c)Natalie Ray/McGraw-Hill Education; **370** (bl)©Corbis/PunchStock, (t)C Squared Studios/Getty Images, (br)HelpingHandPhotos/iStock/Getty Images Plus/Getty Images; **371** (t)©Patrick Bennett/Corbis, (b)©Rolf Schultes/dpa/Corbis; **372-373** Gary Rhijnsburger/Masterfile, Design Pics RF/Getty Images; **373** (b)Natalie Ray/McGraw-Hill Education, (t)McGraw-Hill Education, (c)Michael Scott/McGraw-Hill Education; **374** Thinkstock Images/Getty Images; **375** (r)Glow Images/Punchstock, (l)Glow Images/Punchstock; **376** (b)©Steve Prezant/Corbis, (t)ayagiz/iStock/Getty Images Plus/Getty Images; **377** Photofusion Picture Library/Alamy; **378-379** age fotostock/SuperStock; **379** (tl)Image Source/PunchStock, (b)Image 100/PunchStock, (tr)David Buffington/Getty Images; **380-381** Michelle D. Bridwell/PhotoEdit; **381** (b)Natalie Ray/McGraw-Hill Education, (t)Jacques Cornell/McGraw-Hill Education, (c)McGraw-Hill Education; **382** Natalie Ray/McGraw-Hill Education; **382-383** Natalie Ray/McGraw-Hill Education; **383** McGraw-Hill Education; **384** (b)Jacques Cornell/McGraw-Hill Education, (t)Ken Cavanaugh/McGraw-Hill Education, (c)Natalie Ray/McGraw-Hill Education; **384-385** Natalie Ray/McGraw-Hill Education; **385** Natalie Ray/McGraw-Hill Education; **386** Natalie Ray/McGraw-Hill Education; **387** (r)Natalie Ray/McGraw-Hill Education, (l)Natalie Ray/McGraw-Hill Education; **388-389** Forest Johnson/Masterfile, Patrick Orton/Getty Images; **390-391** Natalie Ray/McGraw-Hill Education; **391** Natalie Ray/McGraw-Hill Education; **392** (b)Natalie Ray/McGraw-Hill Education, (t)Natalie Ray/McGraw-Hill Education; **393** ©Patrick Bennett/Corbis; **394-395** eStock Photo/Alamy; **395** (t)Aqua Image/Alamy, (b)Kader Meguedad/Alamy, (c)John Giustina/Getty Images; **396-397** Nine OK/Photographer's Choice/Getty Images; **397** (t to b)McGraw-Hill Education, (1)Ken Cavanaugh/McGraw-Hill Education, (2)McGraw-Hill Education, (3)McGraw-Hill Education, (4)Natalie Ray/McGraw-Hill Education; **398** (b)Thinkstock/Punchstock, (t)Jack Star/PhotoLink/Getty Images; **399** (br)Golden Gate Images/Alamy, (bl)John W Banagan/The Image Bank/Getty Images, (t)Photodisc/Getty Images; **400** (b)Aqua Image/Alamy, (t)Dynamic Graphics/JupiterImages; **401** (b)Natalie Ray/McGraw-Hill Education, (t)matt matthews/iStock/Getty Images Plus/Getty Images; **402-403** ©Hugh Sitton/Corbis; **403** (b)Natalie Ray/McGraw-Hill Education, (bc)McGraw-Hill Education, (t)Ken Cavanaugh/McGraw-Hill Education, (tc)Jacques Cornell/McGraw-Hill Education; **404** John Giustina/Getty Images; **405** (t)C Squared Studios/Getty Images, (b)©Atlantide Phototravel/Corbis; **406** (tl)Creatas/Punchstock, (bl)ViewStock/Getty Images, (br)©Ted Soqui/Corbis, (tr)Tom Carter/PhotoEdit; **407** David Ponton/Design Pics; **408** (t)Rob Casey/Alamy, (b)Zoonar/R.Henderson/age fotostock, (c)Rick Brady/McGraw-Hill Education; **409** (b)Pat LaCroix/The Image Bank/Getty Images, (t)S. Wanke/PhotoLink/Getty Images; **410-411** Jon Bower/Alamy; **411** (t to b)McGraw-Hill Education, (1)Jacques Cornell/McGraw-Hill Education, (2)Jacques Cornell/McGraw-Hill Education, (3)Jacques Cornell/McGraw-Hill Education, (4)Jacques Cornell/McGraw-Hill Education, (5)Natalie Ray/McGraw-Hill Education; **412** (l)Eyewire Collection/Getty Images, (r)Natalie Ray/McGraw-Hill Education, (cr)Eyewire Collection/Getty Images, (cl)Natalie Ray/McGraw-Hill Education; **413** (b)Natalie Ray/McGraw-Hill Education, (t)©Edith Held/Corbis; **414** (l)©Rudy Sulgan/Corbis, (r)Kader Meguedad/Alamy; **414-415** Natalie Ray/McGraw-Hill Education, Natalie Ray/McGraw-Hill Education, Natalie Ray/McGraw-Hill Education; **415** Mark Steinmetz/McGraw-Hill Education; **416** ©Randy Lincks/Corbis; **417** Nigel Reed/Alamy; **418-419** Richard Cummins/age fotostock; **419** (b)Natalie Ray/McGraw-Hill Education, (t)McGraw-Hill Education, (c)Jacques Cornell/McGraw-Hill Education; **421** (inset)McGraw-Hill Education, (bkgd)McGraw-Hill Education; **422** Rich Legg/iStock/Getty Images Plus/Getty Images; **423** Jacques Cornell/McGraw-Hill Education; **424-425** ©Natalie Fobes/Corbis; **426** Digital Vision Ltd./SuperStock; **427** (inset)©Royalty-Free/Corbis, (bkgd)McGraw-Hill Education, (c)Chris Ryan/age fotostock; **428** (b)Natalie Ray/McGraw-Hill Education, (t)John Giustina/Getty Images; **429** (tr)Kader Meguedad/Alamy, (tl)matt matthews/iStock/Getty Images Plus/

Getty Images, (b)©Edith Held/Corbis; **430** (t)Jerzyworks/Masterfile, (bl)UpperCut Images/Glowimages, (br)©Adam Woolfitt/Corbis, (t)Atsuo WATANABE/Moment/Getty Images; **R01** (bl)McGraw-Hill Education, (bcr)McGraw-Hill Education, (br)McGraw-Hill Education, (bcl)McGraw-Hill Education, (t)Jacques Cornell/McGraw-Hill Education; **R03** (c)McGraw-Hill Education, (b)McGraw-Hill Education, (tl)McGraw-Hill Education, (tc)McGraw-Hill Education; **R04** (t)McGraw-Hill Education, (b)Jacques Cornell/McGraw-Hill Education; **R05** McGraw-Hill Education; **R06** (t)Ryan McVay/Getty Images, (br)McGraw-Hill Education, (bl)Photodisc/Getty Images; **R07** Jules Frazier/Getty Images; **R08** ImageState/PunchStock; **R09** (b)McGraw-Hill Education, (t)McGraw-Hill Education; **R10** (t)McGraw-Hill Education, (b)McGraw-Hill Education; **R11** McGraw-Hill Education; **R13** (t)F.Schussler/PhotoLink/Getty Images, (b)©Brand X Pictures/Corbis; **R14** (l)Purestock/PunchStock, (r)Photodisc Collection/Getty Images; **R15** Stockdisc/PunchStock; **R16** (bl)Richard Hutchings/PhotoEdit, (tr)McGraw-Hill Education, (cr)McGraw-Hill Education, (br)Richard Hutchings/PhotoEdit, (bkgd)Richard Hutchings/PhotoEdit, (cl)McGraw-Hill Education, (tl)McGraw-Hill Education; **R17** (tl)Jack Hollingsworth/Getty Images, (bl)Photodisc/PunchStock, (br)D. Berry/PhotoLink/Getty Images, (tr)David Buffington/Getty Images; **R18** Image Source/PunchStock; **R19** (bc)Natalie Ray/McGraw-Hill Education, (c)Perytskyy/iStock/Getty Images Plus/Getty Images, (t)Nigel J. Dennis/Science Source, (b)Johner Images/Alamy, (tc)©MedioImages/SuperStock; **R20** (bc)Natalie Ray/McGraw-Hill Education, (b)NASA/age fotostock, (tc)Juniors Bildarchiv GmbH/Alamy, (t)Ken Cavanaugh/McGraw-Hill Education, (c)©Punchstock/Corbis; **R21** (tc)Natalie Ray/McGraw-Hill Education, (bc)Jack Star/PhotoLink/Getty Images, (b)©Pixtal/age fotostock, (t)tonda/iStock/Getty Images Plus/Getty Images; **R22** (b)©Image Source/Corbis, (t)©Luis Forra/epa/Corbis, (c)Ariel Skelley/Getty Images, (bc)mychadre77/iStock/Getty Images Plus/Getty Images; **R23** (t to b)ARCO/Reinhard H/age fotostock, (1)©Ingram Publishing/age fotostock, (2)©Fritz Rauschenbach/Corbis, (3)Nigel McCall/Alamy, (4)HelpingHandPhotos/iStock/Getty Images Plus/Getty Images, (5)Spiraling Road Photography/GettyImages, (6)Ken Karp/McGraw-Hill Education, (7)Waltraud Ingerl/Getty Images, (8)Gilbert S. Grant/Science Source; **R24** (t to b)Jules Frazier/Getty Images, (1)Stockdisc(Stockbyte)/Getty Images, (2)©iStockphoto.com/Andyworks, (3)Comstock Images/Punchstock, (4)Nicole Duplaix/National Geographic/Getty Images, (5)Dorling Kindersley/Getty Images; **R25** (6)Evan Sklar/The Image Bank/Getty Images, (5)David R. Frazier Photolibrary, Inc./Alamy, (t to b)McGraw-Hill Education, (1)Diane Diederich/iStock/Getty Images Plus/Getty Images, (2)Dynamic Graphics Group/IT Stock Free/Alamy, (3)©Arthur Morris/Corbis, (4)John Warden/The Image Bank/Getty Images; **R26** (t to b)©Steve Prezant/Corbis, (1)Nigel Cattlin/Alamy, (2)Mark Steinmetz/McGraw-Hill Education, (3)Jill Braaten/McGraw-Hill Education, (4)Comstock/PictureQuest, (5)Theo Allofs/Photodisc/Getty Images; **R27** (t to b)Natalie Ray/McGraw-Hill Education, (1)©DLILLC/Corbis, (2)Jacques Cornell/McGraw-Hill Education, (3)D. Hurst/Alamy, (4)Ken Karp/McGraw-Hill Education; **R28** (t to b)Ken Cavanaugh/McGraw-Hill Education, (1)Dennis Gray/Cole Group/Getty Images, (2)Eckhard Slawik/Science Source, (3)Thinkstock/Punchstock, (4)A & C Wiley/Photolibrary/Getty Images; **R29** (c)Brand X Pictures/PunchStock, (tc)C. Borland/PhotoLink/Getty Images, (bc)Jeff Hunter/Getty Images, (t)©Rainer Hackenberg/Corbis, (b)Eckhard Slawik/Science Source; **R30** (b)Deco/Alamy, (tc)©Don Hammond/age fotostock, (bc)Jacques Cornell/McGraw-Hill Education, (t)Tom Carter/PhotoEdit; **R31** (t to b)Ken Cavanaugh/McGraw-Hill Education, (1)Michael Scott/McGraw-Hill Education, (2)Glow Images/Punchstock, (3)Chris Clinton/The Image Bank/Getty Images, (4)Nolmedrano99/iStock/Getty Images Plus/Getty Images, (5)Pixtal/age fotostock; **R32** (b)IT Stock/Punchstock, (bc)Natalie Ray/McGraw-Hill Education, (t)Photofusion Picture Library/Alamy, (c)BigCheesePhoto/PunchStock, (tc)Photka/iStock/Getty Images Plus/Getty Images; **R33** (t to b)Natalie Ray/McGraw-Hill Education, (1)A & C Wiley/Photolibrary/Getty Images, (2)Maximilian Stock Ltd/Science Source, (3)Westend61 - FotoFealing/Brand X Pictures/Getty Images, (4)Judd Pilossof/Photolibrary/Getty Images; **R34** (t to b)Bogdan Wankowicz/Alamy, (1)Sean Russell/Getty Images, (2)Thinkstock Images/Getty Images, (3)Natalie Ray/McGraw-Hill Education, (4)Ken Karp/McGraw-Hill Education, (5)Digital Vision/Getty Images; **R35** (t to b)©Richard Cummins/Corbis, (1)Marius Hepp/EyeEm/EyeEm/Getty Images, (2)McGraw-Hill Education, (3)Photodisc/PunchStock, (4)©G. Schuster/Corbis, (5)David Boag/Oxford Scientific/Getty Images; **R36** (t to b)Brand X Pictures/PunchStock, (1)Jacques Cornell/McGraw-Hill Education, (2)©Royalty-Free/Corbis, (3)Gary Yeowell/The Image Bank/Getty Images, (4)John Giustina/Getty Images; **R37** (b)Rob Casey/The Image Bank/Getty Images, (bc)Creatas/SuperStock, (tc) Wilfried Krecichwost/Stockbyte/Getty Images, (t)Robert Brenner/PhotoEdit; **R38** (t to b) Rosemary Calvert/Getty Images, (1)Art Wolfe/The Image Bank/Getty Images, (2)Natalie Ray/McGraw-Hill Education, (3)Natalie Ray/McGraw-Hill Education, (4)jaki good photography - celebrating the art of life/Moment Open/Getty Images, (5)©Gary Carter/Corbis, (6)Natalie Ray/McGraw-Hill Education; **R39** (t to b)Natalie Ray/McGraw-Hill Education, (1)Jacques Cornell/McGraw-Hill Education, (2)Shaun Egan/Getty Images, (3)Wirepec/Getty Images, (4)Natalie Ray/McGraw-Hill Education, (5)Natalie Ray/McGraw-Hill Education.